S0-ELC-565

PERMAFROST
OR
PERMANENTLY FROZEN GROUND
AND
RELATED ENGINEERING PROBLEMS

COMPILED BY

SIEMON WILLIAM MULLER

Geologist, U. S. Geological Survey

Professor of Geology, Stanford University

J. W. EDWARDS, INC.—ANN ARBOR, MICHIGAN

1947

Published in 1943 and 1945 by
Military Intelligence Division
Office, Chief of Engineers
U. S. Army

Reprinted by Arrangement with
Office of Technical Services
Department of Commerce

LITHOPRINTED BY EDWARDS BROTHERS, INC.
ANN ARBOR, MICHIGAN, U.S.A.

Preface to Second Printing.

In the second printing of the Strategic Studies No. 62 most of the typographical and other errors which inadvertently crept into the first printing of March, 1943, have been corrected. Several drawings have been improved and some new material has been added. No time was available to make a thorough revision of the text and to organize and to incorporate additional pertinent information on hand.

It is with pleasure that I wish to express my gratitude to Lt. Col. Boyd Yaden, Division Engineer, Alaskan Division, A.T.C., and to Brigadier General Dale V. Gaffney, Commanding General, Alaskan Division, A.T.C., for the opportunity during the past year to study the permafrost phenomena in the field. Both General Gaffney and Colonel Yaden were most helpful in providing the necessary facilities and personnel to aid me in carrying on this investigation. General Gaffney and Col. Yaden have thorough understanding of the importance of the permafrost problem and their deep interest in this project was in no small measure responsible for the results attained to date and for the current participation in this study by other government agencies and by various individuals.

I derived much benefit from the cooperative effort and the help received from the personnel of the office of the St. Paul District Engineer, particularly from Capt. H. A. Humphrey with whom it was my privilege to discuss the various aspects of the problem in the field and in the office.

My associates, officers, enlisted men, and civilian personnel, in the office of the Division Engineer, Alaskan Division, A.T.C., have been most helpful in various ways.

I also acknowledge with pleasure the assistance of many others who have aided me in the preparation of the present text.

SIEMON WM. MULLER
Chief, Special Projects Section
Air Installations Division
Alaskan Div., A.T.C., A.A.F.

C O N T E N T S

LIST OF ILLUSTRATIONS

Page

Page

INTRODUCTION

Permafrost or permanently frozen ground is a widespread phenomenon in northern Asia and in northern North America. It is found in almost one-half of the territory of the U.S.S.R., from the Arctic to northern Mongolia and Manchuria, - an area considerably larger than the entire U.S.A. Permafrost also occurs in most of Alaska and northern Canada. Altogether about one-fifth of all the land area of the world is underlain by permanently frozen ground.

The thickness of permafrost ranges from several hundred meters in the north to a fraction of a meter in the south. The known maximum thickness of about 400 meters was reported from Amderma on the arctic coast of Russia.

Although the existence of permanently frozen ground has been known for a long time, relatively little comprehensive and systematic work had been done in this field until about two decades ago when scores of Russian scientists began an intensive research which already has brought fruitful results.

The destructive action of permafrost phenomena has materially impeded the colonization and development of extensive and potentially rich areas in the north. Roads, railroads, bridges, houses, and factories have suffered deformation, at times beyond repair, because the condition of frozen ground was not examined beforehand, and because the behavior of frozen ground was little if at all understood. In consequence the structures were built without the proper protective features that would enable them to withstand the subsequent heaving, settling, and other effects of frost action.

Stresses that develop in freezing ground may exceed 2,000 kilograms per square centimeter. To meet such stresses by structural design alone is highly uneconomical, if feasible at all. Therefore, it is clear that the usual working methods and practices cannot be successfully employed in the permafrost region unless they are radically modified to meet the unique permafrost conditions.

Costly experience of Russian engineers has shown it to be a losing battle to fight the forces of frozen ground simply by using stronger materials or by resorting to more rigid designs. On the other hand, this same experience has demonstrated that satisfactory results can be achieved if the dynamic stresses of frozen ground are carefully analyzed and are allowed for in the design in such a manner that they appreciably minimize or completely neutralize and eliminate the destructive effect of frost action. Mastery of this working principle, however, can be achieved only if the natural phenomena of frozen ground are thoroughly understood and their forces are correctly evaluated. In the far North a comprehensive and systematic study of frozen ground, therefore, should constitute an integral part of the planning and design of any engineering project.

Judging from the similarity of conditions in Siberia and Alaska it should not come as a surprise if the newly built Alcan Highway suffers certain damage causing delays and even disruption of traffic The most destructive swelling of ground and heaving of structures usually occur in the winter and are followed by almost equally destructive settling, caving, and sloughing of ground during the summer thaw. Unfortunately, as far as is known, no systematic observations on the frozen ground phenomena were made during the construction of the Alcan Highway. Such studies, had they been made, would have been of immeasurable value in making repairs of the damage that is likely to occur during the first winter, spring, and summer and in the more distant future.

In this connection, it is worth noting that in Soviet Russia since about 1938 all government organizations, municipalities, and cooperative societies are required to make a thorough survey of the permafrost conditions according to a prescribed plan before any structure may be erected in the permafrost region.

The purpose of this report is to acquaint those unfamiliar with the Russian language with the scope of the problem, the progress made in this field of investigation during recent years, and to summarize the results that have direct practical application to various engineering problems.

The present compilation is based for the most part, on articles by Sumgin, Tsytovich, Bykov, Tolstikhin, and Lukashev, but draws also upon many other foreign and American sources. A selected list of references is given in the bibliography on pages 130-135.

Technical terminology pertaining to the frozen ground phenomena described in the present report is, for the most part, new and represents either a more or less free translation or a direct adoption of terms widely used in Russia. The terms with their explanations are listed in the glossary at the end of the article.

I am deeply indebted to my colleagues at the United States Geological Survey, particularly to W. H. Bradley, H. G. Ferguson, J. T. Hack , and P. S. Smith who have read the manuscript and offered many helpful suggestions.

I also had an opportunity on several occasions to discuss the subject of this report with Professor K. I. Lukashev of the Leningrad University and Dr. C. C. Nikiforoff of the U. S. Department of Agriculture, both of whom studies frozen ground phenomena in Siberia.

I have derived much benefit from these discussions in which both Nikiforoff and Lukashev have shared with me much of the unpublished data based on their field observations.

Miss Marie Siegrist of the Geological Society of America and Miss Z. I. Gruber of the Geological Survey have been very helpful in the search for the references and in obtaining books and periodicals from the Library of Congress and from the Library of the U. S. Geological Survey.

PERMAFROST OR PERMANENTLY FROZEN GROUND

Definitions

Permanently frozen ground or permafrost is defined as a thickness of soil or other superficial deposit, or even of bedrock, at a variable depth beneath the surface of the earth in which a temperature below freezing has existed continually for a long time (from two to tens of thousands of years). Permanently frozen ground is defined exclusively on the basis of temperature, irrespective of texture, degree of induration, water content, or lithologic character.

The phenomenon of permanently frozen ground is also known as "permanently frozen soil" or "ever frozen soil", but it is believed that the expression "permanently frozen ground" is the most appropriate, particularly as the permanently frozen condition commonly extends well below the level of soil and in many cases affects even a more or less solid bedrock.

The expression "permanently frozen ground", however, is too long and cumbersome and for this reason a shorter term "permafrost" is proposed as an alternative.

Ground with temperature below 0° C.[1/] but containing no ice as cementing substance is called dry frozen ground and if it is permanent it is called dry permanently frozen ground or dry permafrost.

Dry frozen condition is usually found in sandy or in other coarse-grained clastic material that drains easily. From the standpoint of construction engineering the properties of dry frozen ground are similar to those of unfrozen ground. In the grading or excavating operations dry frozen ground has neither the hardness nor the induration which are characteristic of permanently frozen ground that contains moisture in the form of ice. Nonetheless, in some engineering projects, such as sewage disposal and other pipe-line construction, dry frozen ground cannot be ignored for, despite the ease of excavation, the sub-zero temperature may freeze and damage the pipes.

1/ Centigrade temperatures and metric system units are used throughout this text. Conversion table, metric to English units, see p. 231.

Bykov, Kapterev, and Lukashev distinguish between permanently frozen ground and simply frozen ground. According to the generally accepted definition, permanently frozen ground may contain little or no water but frozen ground always implies the presence of water in the form of ice. It should be noted in this connection that not all ground that contains water and whose temperature is below 0° C. is necessarily frozen ground, for there are conditions under which water does not turn into ice even at temperatures considerably below 0° C.

Ground frozen by a sudden drop of temperature and remaining frozen but a short time, usually a matter of hours, is referred to as transitory frozen ground. Ground frozen for a longer time by low winter temperatures is known as seasonally frozen ground.

Origin of permafrost

Permafrost can exist, and therefore could have originated, only where the mean annual temperature is below 0° C. There are still some differences of opinion as to exactly what minimum temperature is required. Some evidence has been offered to support the contention that the mean annual temperature cannot be higher than -3.3°, -4°, or even -6° C. but Sumgin shows that, although it has to be below the normal freezing point, it can, nevertheless, be quite close to 0° C.

It is now generally agreed by most authors that permafrost first appeared during the refrigeration of the Earth's surface at the beginning of the Pleistocene or Ice Age, perhaps a million years ago. It is also generally believed that during the subsequent periods of climatic fluctuations corresponding changes must have taken place in the thickness and areal extent of permafrost.

Although the relict nature of permafrost appears to be established, it is equally certain that, at least in some areas, permafrost is forming today under the present climate. Such is the case, for example, in some of the recently built river islands and bars in the Arctic part of Siberia.

A suggestion that a relationship may exist between the areal extent of the Pleistocene glaciation and the present geographic distribution of permafrost has been advanced by Nikiforoff and more recently by Gerasimov and Markov. The available data on this question are as yet too meagre to permit any definite conclusions but the cumulative evidence seems to indicate that unusually thick permafrost tends to occur in non-glaciated areas as, for example, in northeastern Siberia. (See Fig. 1).

Geography of permafrost

Permafrost underlies about one-fifth of the entire land surface of the world. It is most widespread in the northern hemisphere around the shore of the Arctic Ocean (See Fig. 1) but is also extensive in the Antarctic.

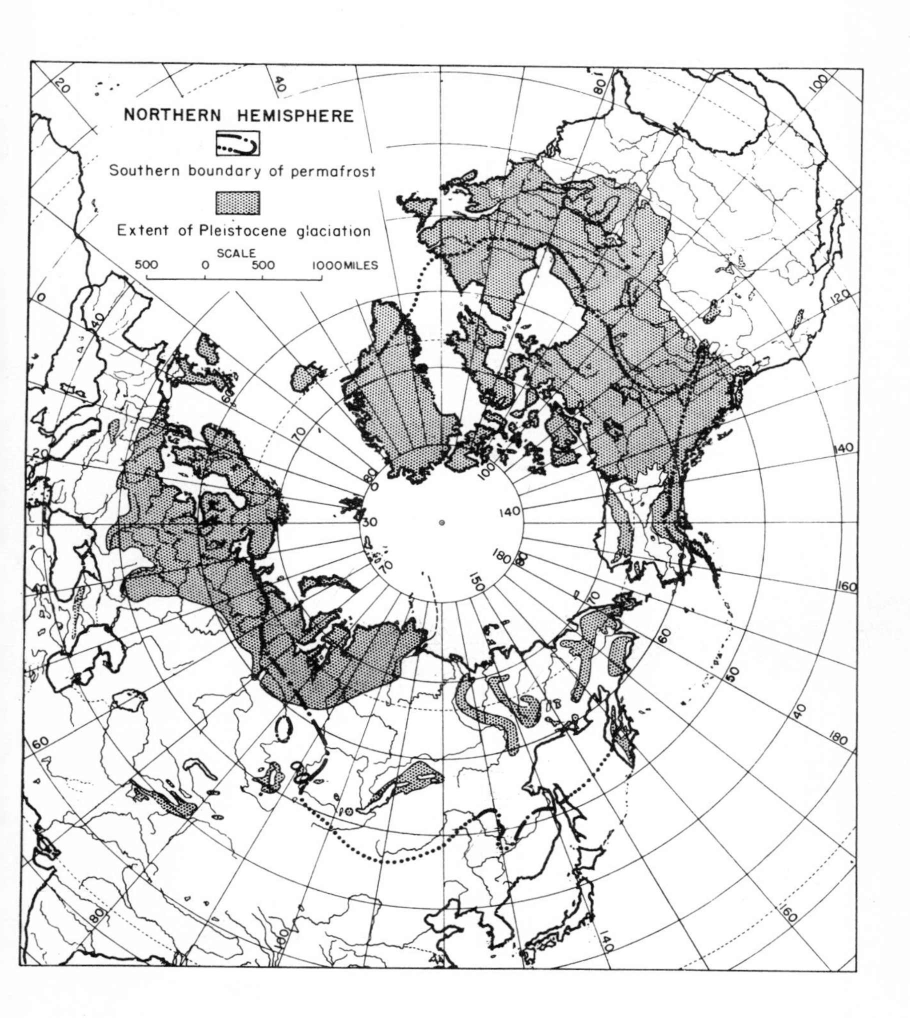
NORTHERN HEMISPHERE
Southern boundary of permafrost
Extent of Pleistocene glaciation
SCALE
500 0 500 1000 MILES

FIG. 1

In North America, according to Nikiforoff, the boundary of the permafrost area roughly follows the course of the Yukon River then runs easterly along the 60th parallel to the 110th meridian where it swings southeast toward the south end of Hudson Bay. From Hudson Bay the boundary line turns northeastward and reaches the Atlantic Ocean on the east coast of Greenland, opposite Iceland.

Bratsev draws the permafrost boundary in western North America about half way between the Yukon River and the Pacific Coast and shows a narrow tongue of permafrost along the Rocky Mountains extending southward to the U. S. boundary. P. S. Smith of the U. S. Geological Survey, in an unpublished manuscript, presents evidence to show that the permafrost in Alaska extends even farther south, almost reaching the coastal region and roughly coinciding with the -1° C. (30° F.) isotherm.

Outside of the main areas outlined above there are islands of permafrost, _sporadic permafrost_, in other parts of the world, usually at altitudes above the snow-line.

In Eurasia the following subdivisions of the permafrost area are recognized by Sumgin: (See Fig. 2)

1. Areas of continuous permafrost wherein the temperature of the ground at the depth of 10-15 meters is prevailingly below -5° C.
2. Area of permafrost having islands of unfrozen ground (_taliks_). Temperature of the ground at a depth of 10-15 meters is prevailingly between -5° and -1.5° C.
3. Areas of sporadic permafrost consisting dominantly of unfrozen ground but containing islands of permafrost. Temperature at a depth of 10-15 meters is, as a rule, above -1.5° C.
4. Isolated small areas, as in the Kola Peninsula, where permafrost is found only in peat mounds.

Thicknesses of Frozen Ground

The thickness or the depth of penetration of transitory frozen ground is usually measured in millimeters or centimeters. The thickness of permafrost is commonly several meters but in many places, especially along the coast of the Arctic Ocean, it exceeds one or two hundred meters. At Amderma, in northern Russia, permafrost was penetrated to a depth of 230 meters and its thermal gradient indicated that it probably extended downward 120 meters farther. (See Fig. 3)

Above the permafrost is a layer of ground that thaws in the summer and freezes again in the winter. This layer represents the seasonally frozen ground and is called the _active layer_.

The thickness of the active layer is variable. As a rule it is fairly thin in the north and becomes thicker to the south. Its thickness also depends on the composition of the ground and vegetation.

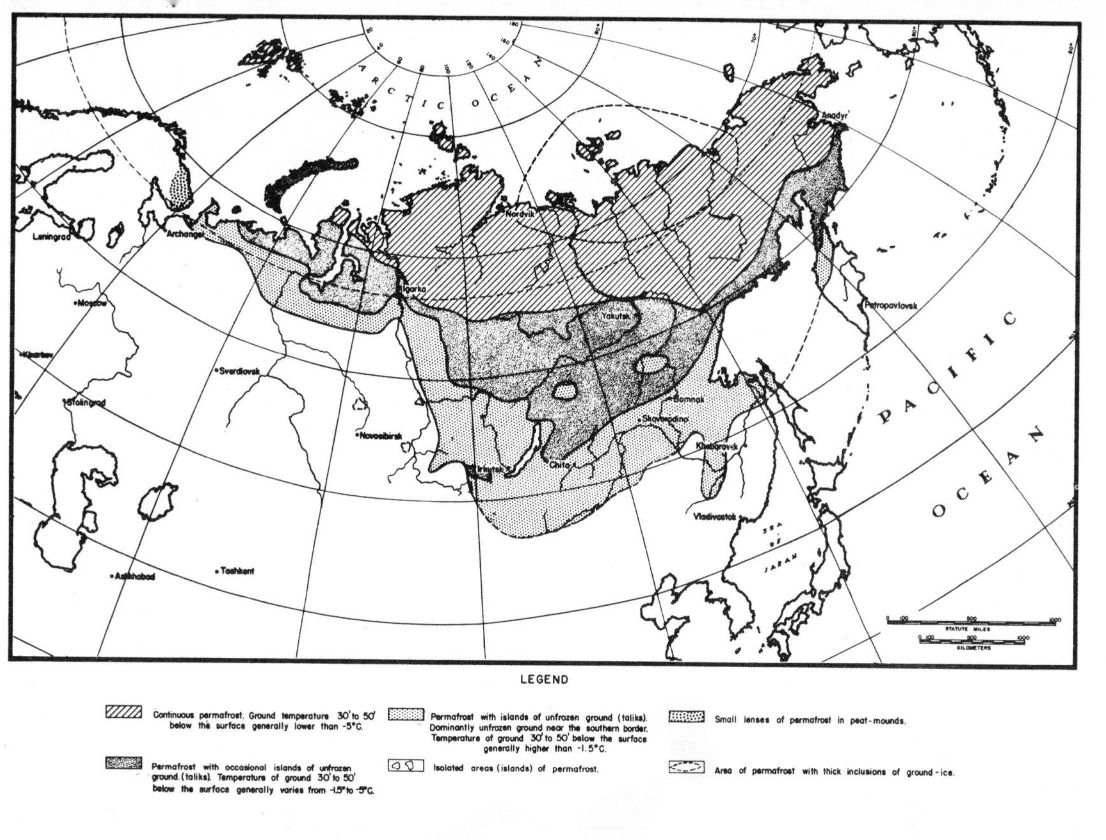

FIG. 2

(After Sumgin)

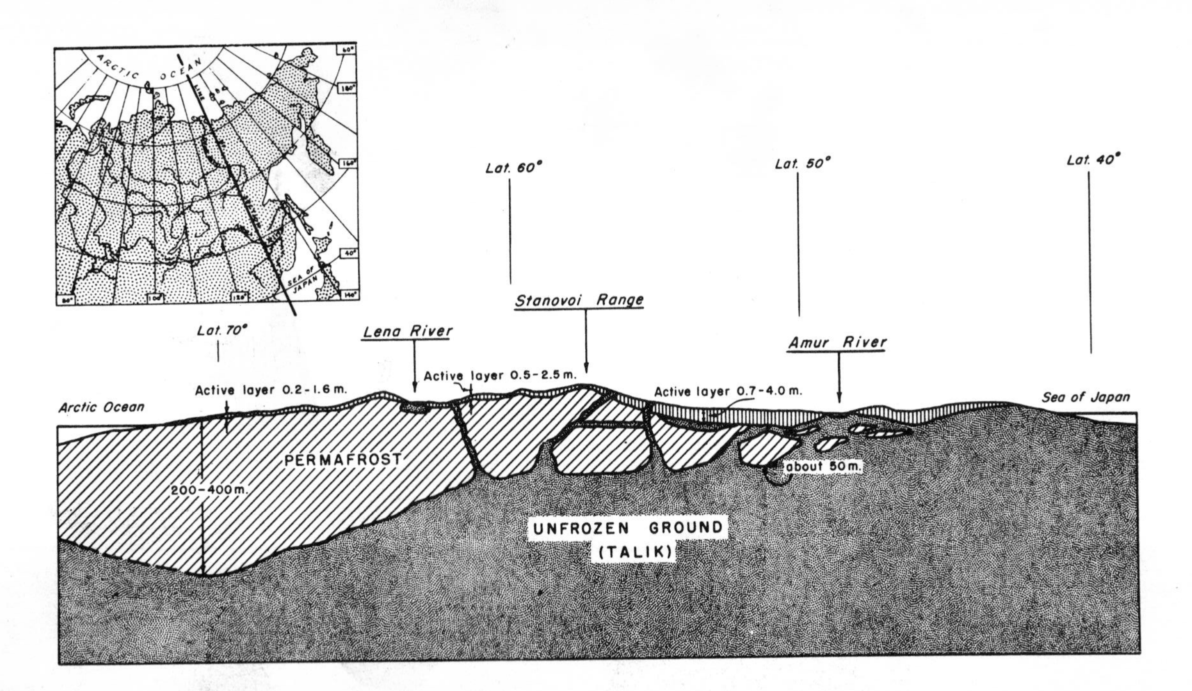

Diagrammatic cross-section through Siberia, from the Arctic Ocean to the Sea of Japan, showing relative thickness of permafrost and active layer.

FIG. 3

In Siberia the average thickness in meters of the active layer in different types of ground and in different surroundings is as follows:

	In sandy ground	In clayey ground	In ground with peat and swamps
In areas south of the 55th parallel	3 to 4	1.8 to 2.5	0.7 to 1.0
At the latitude of Yakutsk (62° N.)	2 to 2.5	1.5 to 2.0	0.5±
Along the coast of the Arctic Ocean	1.2 to 1.6	0.7 to 1.0	0.2 to 0.4

Other conditions that affect the thickness of the active layer are type of surface exposure, hydrology, vegetation and snow cover.

The local variation in the thickness of the active layer is illustrated by the following example taken from the southeastern part of the permafrost area in Siberia.

On a river terrace composed of water-saturated sandy clay covered by 0.5 meter of peat and moss and having a sparse stand of larch, the thickness of the active layer is 0.5 to 0.8 meter.

Under essentially the same conditions except that instead of larch the vegetation consists of grasses and stands of broadleaf trees, such as birch, alder, and poplar, the thickness of the active layer is 1.5 to 2.5 meters.

In the same valley, on the next higher terrace, which is composed of well drained sandy material and has a vegetation cover consisting of lichens, dry mosses, and tall pines, the thickness of the active layer is likely to be between 2.5 to 3.5 meters.

Similar relations hold true for the more northerly areas in the permafrost region although the actual thicknesses are somewhat smaller.

The rate and depth of seasonal thawing are also markedly influenced by the type of exposure, hydrology, and vegetation and snow cover. An insulating cover of snow or vegetation will cause a shallower thaw as contrasted with a deeper thaw in places that are bare or that receive more sunshine. Owing to the fact that these conditions vary considerably within short distances the bottom of the thawed ground is a rather uneven surface. By analogy with ground-water terminology this surface will be called the frost table. With the downward progress of the seasonal thaw the frost table moves progressively lower until it either disappears when it reaches underlying unfrozen ground or merges with the upper surface of the permafrost.

The surface that represents the upper boundary of permafrost, may be called the permafrost table. It does not necessarily coincide with the frost table. Like the frost table, however, its irregularities are determined by the insulating cover of vegetation, by the difference in the heat conductivity of the ground, by the geographic position and the character of exposure, and by the hydrology of the ground. For example, in the sketch reproduced from Nikiforoff (Fig. 4) the permafrost table is higher where the ground is insulated by the cover of peat and moss and is slightly depressed where the ground is bare.

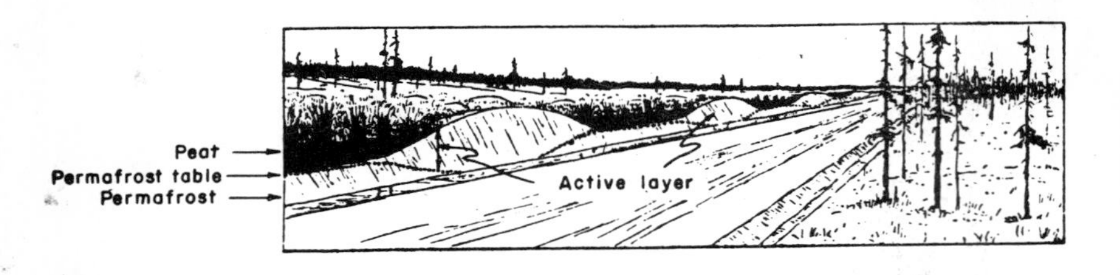

Permafrost profile in a swamp, Amur Province, Siberia
(From Nikiforoff)

FIG. 4

Locally, especially near the southern margin of the permafrost province, the permafrost table lies at a depth that is greater than the level reached by the winter freezing and as a result a layer of unfrozen ground occurs between the frozen active layer above and the permafrost below. This layer of unfrozen ground is designated by the Russian term talik, meaning thawed ground. The term talik is also applied to layers or lenses of unfrozen ground that may occur within the permafrost and to the unfrozen ground that lies beneath the permafrost. Permafrost with the intercalations of taliks is spoken of as the layered permafrost. Individual taliks in a layered permafrost commonly serve as acquifers and their water is, as a rule, under considerable hydrostatic pressure. Taliks, therefore, may be useful as a source of water supply and also as suitable beds in which to lay water pipes. The pipes imbedded in a talik are not likely to freeze and break. Occasionally some of these taliks, particularly those composed of silty material, behave as more or less viscous liquids. They flow like molasses and are comparable to mud-flows but they may occur within a wide range of depths below the surface and are not restricted to the superficial material. This fluid muddy material is designated by the term slud, a provincial English word for soft, wet mud or mire. The corresponding Russian term is "pluvoon", which means "that which flows".

Where permafrost is relatively thin and the overlying active layer consists of porous material the relatively warm surface water may percolate freely through this porous ground and thaw its way clear through the underlying permafrost. Such islands of unfrozen ground, completely surrounded by permafrost, are also called taliks. It is likely that some of these islands of talik have formed where the permafrost was either dry or only partly saturated with ice thus permitting a free percolation of relatively warm surface water. Islands of talik become progressively more common toward the southern boundary of the permafrost region.

The lower limit or the base of the active layer differs in position slightly from year to year. Occasionally, owing to excessive winter cold or to subnormal summer temperatures, the summer thaw stops short of the usual level, leaving a layer of frozen ground between the thawed part of the active layer and the talik. This isolated frozen layer, lasting only a season or two, is designated by the Russian term _pereletok_, meaning "that which survived over the summer". (See Fig. 5).

The combined thickness of ground above the permafrost consisting of the active layer and talik, and, wherever present, the pereletok, is referred to as the _suprapermafrost layer_ or zone.

The thickness of the suprapermafrost layer depends chiefly on hydrothermal conditions of the place and to a lesser extent on the latitude. Near the southern limit of the permafrost province this layer is generally thicker than the active layer. This relationship may also exist locally farther north, especially along the slopes of mountains and near large rivers and lakes. But in the greater part of the permafrost province the suprapermafrost coincides with the active layer.

In most engineering construction where the suprapermafrost layer is fairly thick, 20 meters or more, only the active layer need be considered in a design of shallow foundations. On the other hand, where the suprapermafrost layer is less than 5 meters thick in designing a building the preliminary investigation should include not only the behavior of the active layer but the general effect of permafrost on the hydrologic regime of the overlying ground as well.

Climatic Effects and Thermal Regimes of Permafrost

In general, the following conditions appear to be favorable for the existence of permafrost:

1. Cold and long winter with little snow.
2. Short, dry, and relatively cool summer.
3. Small precipitation during all seasons.
4. Areas with impeded surface and subsurface drainage.

Available meteorological data are as yet inadequate to express these conditions in a quantitative form.

The temperature of permafrost shows relatively small variation with depth. As is shown in Fig. 6 only the upper part of permafrost is affected by the seasonal variation of temperature.

The range or amplitude of this seasonal variation diminishes downward and comes to zero at a level which the Russians call the level of _zero annual amplitude_ or, in short, the level of _zero amplitude_. This

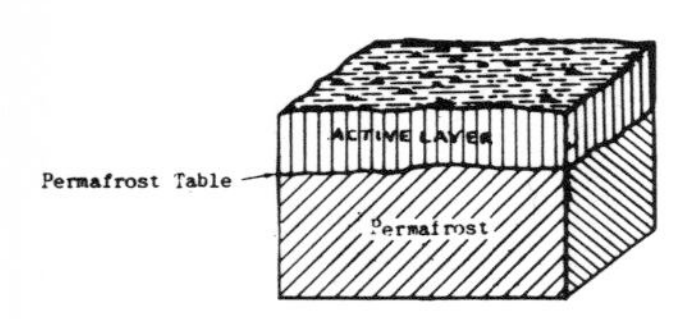

1. Active Layer Extending to Permafrost Table.

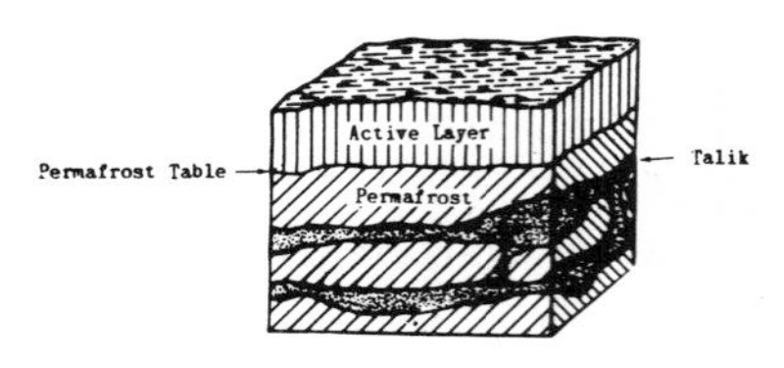

2. Layers of Talik in Permafrost.

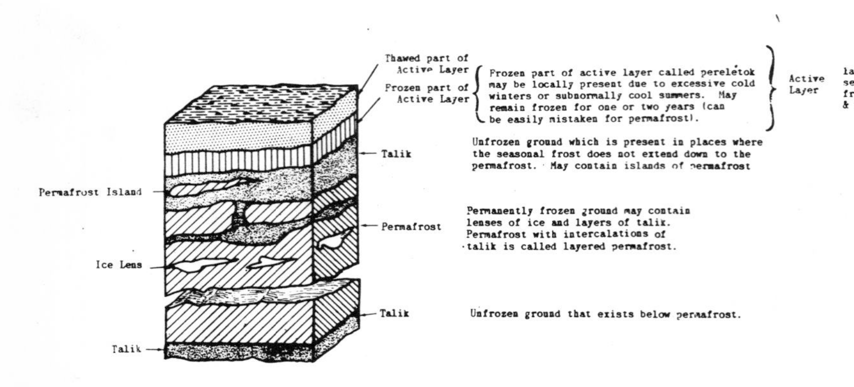

3. General Block Diagram Showing Different Layers in Ground That May Occur in the Permafrost Region.

FIG. 5

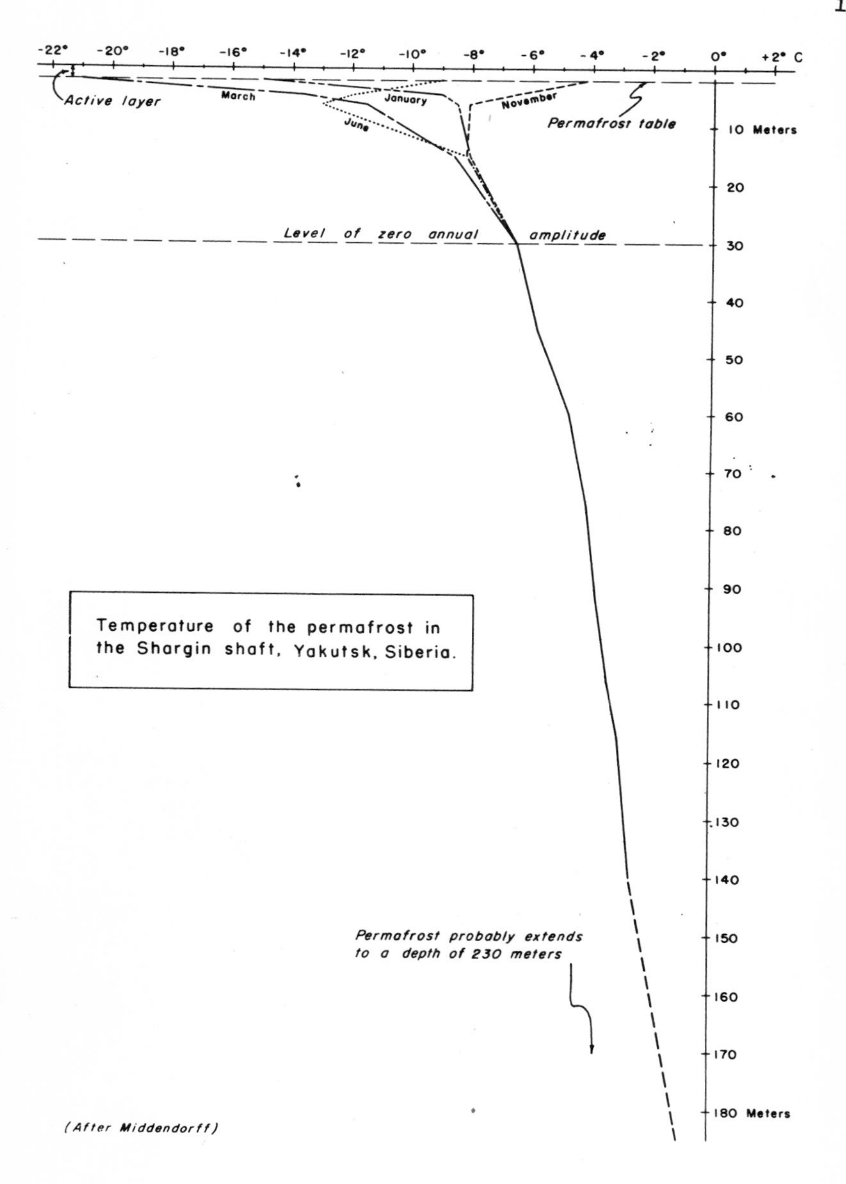
-22°
-20°
-18°
-16°
-14°
-12°
-10°
-8°
-6°
-4°
-2°
0°
+2° C
Active layer
March
January
November
June
Permafrost table
10 Meters
20
Level of zero annual amplitude
30
40
50
60
70
80
90
100
110
120
130
140
150
160
170
180 Meters
Temperature of the permafrost in the Shargin shaft, Yakutsk, Siberia.
Permafrost probably extends to a depth of 230 meters
(After Middendorff)

FIG. 6

horizon is also known as the depth of seasonal change. Below the level of zero amplitude the temperature of the permafrost has a gradient which remains stable from season to season and from year to year. The temperature may continually rise with depth until the unfrozen ground is reached below the permafrost or it may first show a decline and then a rise as is illustrated in Fig. 7.

The greatest fluctuation of temperature takes place in the active layer and is the main cause of damage to buildings, excavations, pipelines, etc.

Temperature and distribution of heat in the ground are controlled by many factors, chiefly climatologic. These are divided into two groups: Constant and Variable.

Constant factors include:

1. Geographic position (latitude, exposure to sun, etc.)
2. Relief
3. Temperature of air
4. Cloudiness
5. Precipitation
6. Direction of prevailing winds

Variable factors are:

1. Snow cover
2. Vegetation
3. Moisture content of ground
4. Heat conductivity of ground
5. Surface evaporation

Practical distinction between these two groups of factors lies in the fact that those of the first group tend to stabilize local climatic conditions and, if correctly evaluated, provide means for computing building specifications. The factors of the second group, however, may to a certain extent be controlled by man and can be employed to change the thermal balance of the ground in a desired direction.

Buildings and earth fills in the permafrost area tend to alter the local thermal regime of the ground causing the permafrost table to change its position. (Figs. 8 and 9)

The new thermal regime becomes nearly stable in a very short time, usually within a year. Occasionally, however, this is not so. For example, a speedily constructed large fill composed of unfrozen ground, containing certain amount of stored heat, may cause a temporary sag in the permafrost table beneath the fill. This sag will ultimately disappear and the permafrost table will become stabilized at a higher level, usually invading the lower part of the newly made fill.

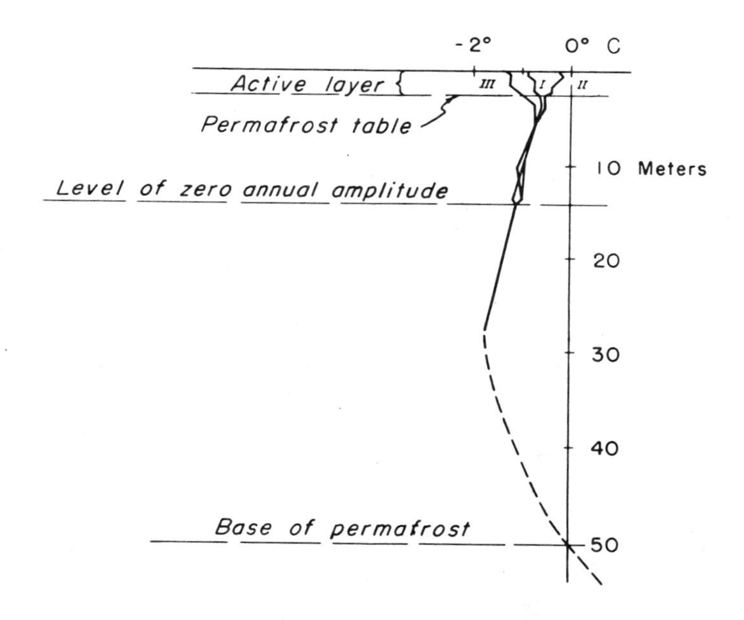

Temperature gradient in the permafrost at Skovorodino, Siberia.

I - Mean temperatures for 1928-1930
II - Temperatures for 1930
III - Temperatures for 1929

(From Sumgin)

FIG. 7

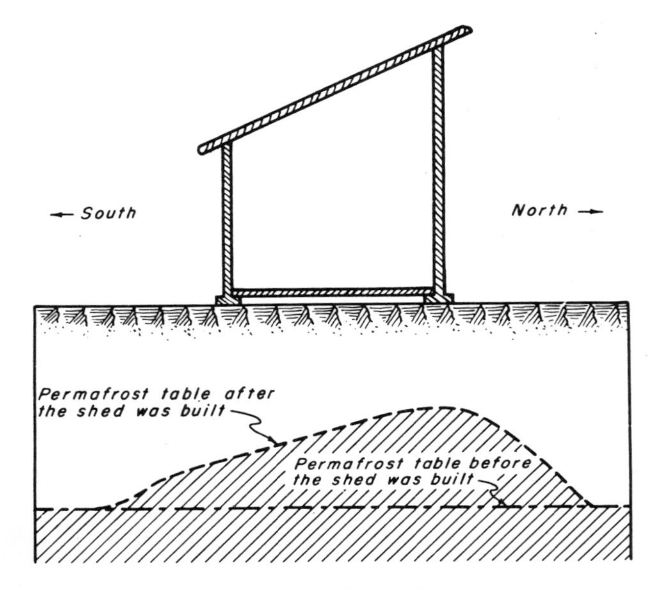

The effect of uninhabited (unheated) building on permafrost table.

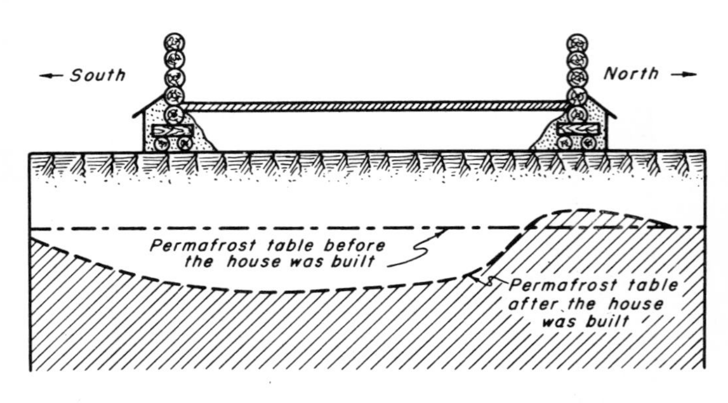

The effect of inhabited (heated) house on permafrost table.

(From Tsytovich and Sumgin)

FIG. 8

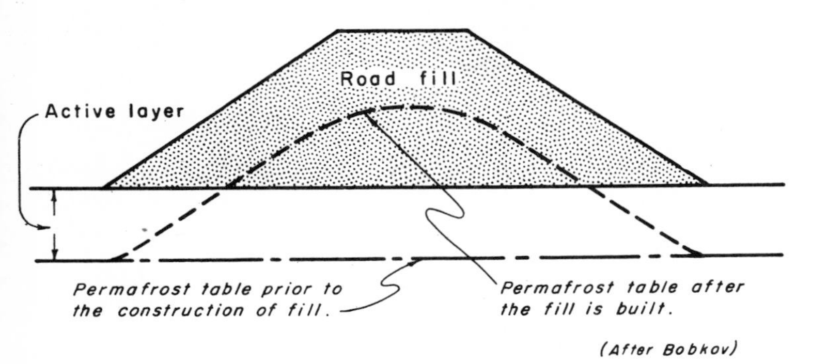

FIG. 9

In all construction work in the permafrost area it is very important to consider the time of the year during which the work is done. A fill constructed in late summer or early fall will contain a considerable amount of stored heat, which will retard the stabilization of thermal regime.

Some students of frozen ground are of the opinion that once the thermal regime is adjusted to the newly erected structures it remains fairly stable and shows little or no tendency to change. Tsytovich and Sumgin, on the other hand, maintain that after a structure is erected the ground underneath will for many years continue to show slight changes in its thermal regime unless appropriate measures are taken to prevent the change.

During the winter freezing the lowering of temperature in moist ground does not proceed downward at a uniform rate. Layers of ground with much moisture retard the downward penetration of cold and remain for a considerable time at about 0° C. while adjacent ground is being chilled to a much lower temperature. In such wet layers the loss of heat is temporarily compensated by the latent heat of fusion that is given off by the water until all of it turns into ice. This condition is called by Sumgin the zero curtain. (Fig.10). The zero curtain also occurs in the ice-saturated layers during the summer thaw when a certain amount of heat is absorbed during the melting of ice.

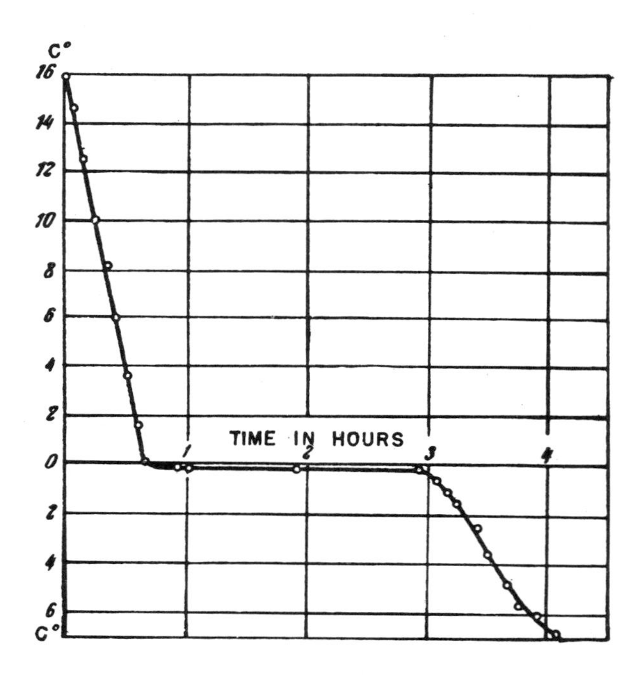

Temperature curve of freezing ground as a function of time.

The break in slope along the 0° line represents the ZERO CURTAIN - an interval during which the temperature remains constant as the chilling of water is temporarily compensated by the latent heat of fusion.

(From Tsytovich and Sumgin)

FIG. 10

Near the base of the active layer where there is usually a considerable concentration of water which is unable to escape downward through the impervious permafrost the zero curtain normally lasts more than a month and in exceptional cases may persist for as many as 115 days in a single year.

The depth at which the zero curtain is likely to occur is controlled by the following factors:

1. Degree of saturation of the ground.
2. The heat conductivity of the ground.
3. Insulating effect of vegetation.
4. Insulating effect of snow.
5. Other factors such as latitude, mean temperature of the air, exposure to sun, etc.

The practical significance of the zero curtain lies in the fact that it permits an artificial regulation of the heat balance in the ground as applied to water-tight fills that are built around the water pipes. This is discussed under the engineering problems in the latter part of this report.

Mean monthly temperatures of the ground recorded at Siberian permafrost stations (See table below.) show that in the active layer 1.5 meters below the surface the minimum monthly temperature occurs in March, whereas the minimum temperature of the air is recorded in January. There is thus a lag of two months. The maximum temperature in the ground occurs in September, also showing a two months' lag.

MEAN MONTHLY TEMPERATURES OF GROUND AT BOMNAK, SIBERIA (in C°)

Depth in meters	Period of Observation	Jan.	Feb.	Mar.	Apr.	May	June	July	Aug.	Sept.	Oct.	Nov.	Dec.	Ann.	
1.5	1910-1919	-0.8	-2.6	-3.8	-2.9	-1.3	-0.7	-0.3	0.8	1.6	0.7	0.1	0.0	-0.8	Active layer
2.	1910-1919	-0.1	-1.2	-2.4	-2.5	-1.4	-0.9	-0.6	-0.2	0.3	0.2	0.05	0.0	-0.7	
2.8	1911-1919	-0.2	-0.5	-1.2	-1.6	-1.4	-1.0	-0.7	-0.6	-0.4	-0.3	-0.2	-0.2	-0.7	Permafrost

MEAN MONTHLY TEMPERATURES OF GROUND AT SKOVORODINO, SIBERIA (in C°)

Depth in meters	Period of Observation	Jan.	Feb.	Mar.	Apr.	May	June	July	Aug.	Sept.	Oct.	Nov.	Dec.	Ann.	
0.4	1928-1930	-12.9	-13.0	-9.1	-3.3	0.3	6.2	10.7	11.8	7.4	1.5	-2.2	-8.6	-0.9	Active layer
0.8	1928-1930	-8.3	-9.6	-7.6	-3.5	-0.9	1.6	6.4	8.5	6.4	1.6	0.0	-3.4	-0.8	
1.6	1928-1930	-0.9	-3.5	-4.3	-2.9	-1.4	-0.9	-0.4	-1.3	2.4	1.0	0.1	0.0	-0.8	
2.0	1928-1930	-0.4	-1.9	-3.2	-2.6	-1.5	-1.0	-0.7	0.0	1.0	0.6	0.1	0.0	-0.8	
2.5	1928-1930	-0.2	-0.2	-1.0	-1.7	-1.3	-1.0	-0.8	-0.7	-0.4	-0.2	-0.2	-0.2	-0.6	Permafrost

(For conversion of C° into F° see table of measures on the last page)

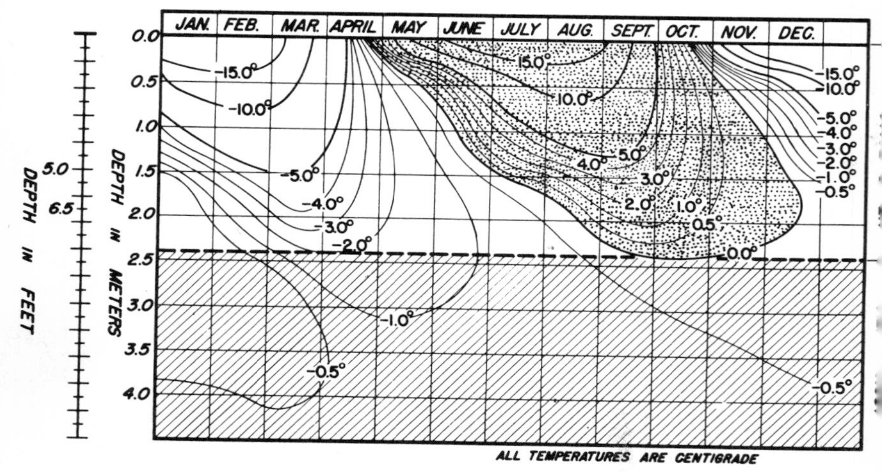

THERMAL REGIME OF GROUND AT SKOVORODINO, SIBERIA (1928—1930)

FIG. 11

At the depth of 2 meters the minimum temperature lags three months, but the maximum, just as at the depth of 1.5 meters, occurs two months later than at the surface.

At the depth of 1.5 meters the change of temperature from negative (below 0° C.) to positive (above 0° C.) takes place in August and the change from positive to negative occurs only in January whereas in December the mean monthly temperature remains at 0° C. and the ground is in the process of freezing.

At 2 meters below the surface the change of temperature from negative to positive occurs in September, one month later than at the 1.5 meter depth but the change from positive to negative takes place in January with December temperature remaining at 0° C., just as it is at the 1.5 meters depth.

From the above it follows that while the air temperatures are the lowest in December and January, the ground at 1.5 and 2 meters below the surface at that time is just beginning to freeze. With further downward penetration of frost the ground-water trapped between the frozen surface ground above and the permafrost below may be subjected to a strong hydrostatic pressure. Swelling of soil and a pronounced heaving of the ground surface generally occur from December on, attaining their greatest intensity in February or March at which time the active layer is frozen down to the maximum depth. (Fig. 11)

Degradation and Aggradation of Permafrost

By degradation of permafrost is meant its disappearance or waning at a given locality owing to a progressive thawing from year to year. Degradation of permafrost may proceed from the top downward by a progressive lowering of the permafrost table or the ground may thaw from the base up, or both. Degradation of permafrost may be caused by:

1. Solar heat.
2. Heat radiation of the Earth.
3. Geologic changes (physico-chemical and biologic changes in the ground).
4. Shifting of surface and subsurface drainage (warming effect of percolating ground water).
5. Buildings and other engineering constructions that change the existing conditions.

Permafrost degradation deserves most careful consideration as the changes in the ground resulting either from natural or artificial causes are likely to affect the stability of a building or other structure. In engineering work, the thermal regime of the ground should be carefully investigated by test drill surveys to a depth considerably greater than may at first appear to be significant. For most projects the ground is usually tested to a depth of 5 to 10 meters. The permafrost table is frequently encountered at a depth of 5 meters below the surface whereas the 10 meter level may be only a layer of zero annual amplitude. It is clear that these data are not sufficient to determine the heat (or cold) balance at deeper levels, say 15 or 20 meters, and do not reveal the trend or gradient of the sub-zero temperatures below the 10 meter level. Furthermore, the degradation of permafrost may proceed more actively from the bottom up than from the top downward and unless the thermal gradient of the ground is ascertained and properly allowed for, the apparent stability of ground may unexpectedly become disturbed, and may cause settling or caving.

In some areas, instead of disappearing, permafrost may actually be spreading or appearing for the first time. This spread or appearance of permanently frozen ground is called aggradation of permafrost.

Aggradation of permafrost under natural conditions has been recorded in rapidly growing deltas of large rivers emptying into the Arctic Ocean and in floodplains where the silting of the valley floor during floods creates a protective blanket over the ground which has not thawed out completely from the preceding winter. In general, however, there is as yet very little factual information available on degradation and aggradation of permafrost.

Hydrology of Frozen Ground

Hydrology of frozen ground takes into consideration the amount of surface and ground water, the direction and rate of flow or percolation, the source of water, and the changes of phases of water from liquid to solid (freezing) and from solid to liquid (melting).

A part of the surface water runs off and feeds brooks, rivers, and lakes, but a considerable proportion of it filters into the ground through the porous surface material and becomes the ground-water. Coarse bouldery talus along the mountain slopes is an effective intake of rain and snow and deserves particular attention in a hydrologic study of a mountainous area. The different paths of percolation of ground water are shown diagrammatically in Fig. 12.

The flow of rivers in permafrost regions is controlled by the nature of their dominant source of water. Most streams in these regions show a prominent rise in stage during the late spring and early summer thaw of snow and ice. Rivers that depend entirely on melting snow fluctuate a great deal and many cease to flow after the snow has melted. Rivers, like the Amur River, depend primarily on summer rains and their peak stage, accordingly, is reached late in the summer, immediately after the rainy season. Many small streams are fed by seepages from the thawed surface ground, the active layer. Streams that depend to a considerable extent on deep springs or on water from cavernous underground channels usually maintain a fairly uniform flow throughout the greater part, if not the entire year.

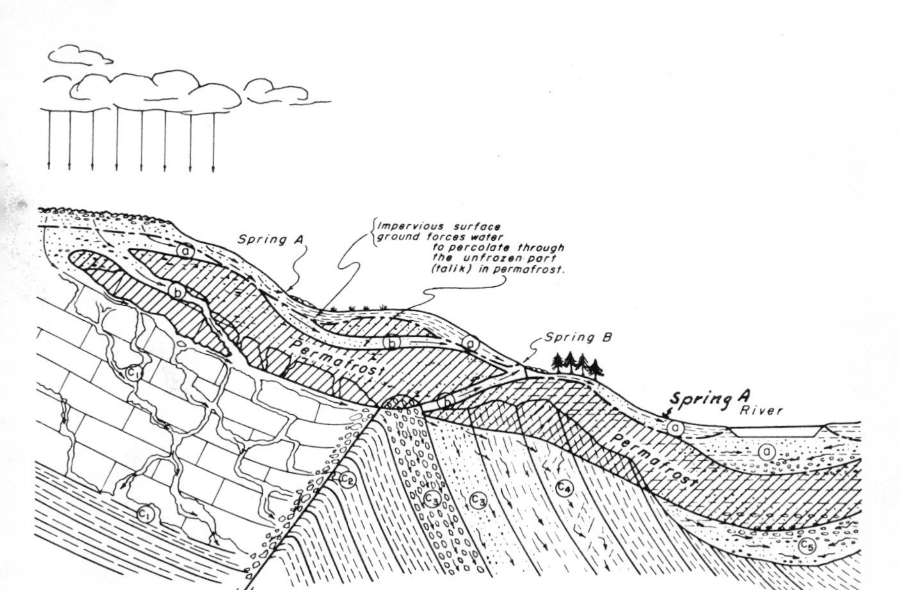

DIAGRAM SHOWING DIFFERENT OCCURRENCES OF GROUND-WATER IN THE PERMAFROST REGION

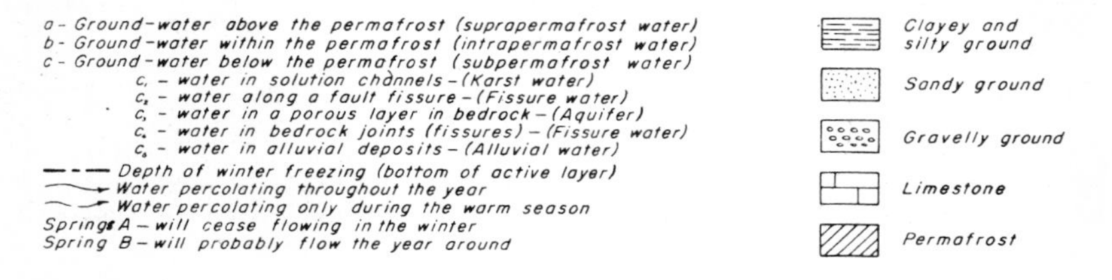

FIG. 12

Large rivers, especially those that do not freeze solid during the winter, have a marked warming effect on the underlying river bed and as a result the ground beneath them remains unfrozen and commonly contains freely percolating water. (Figs. 13 and 14)

Even beneath streams that do freeze clear through the ground may remain unfrozen and may contain water-bearing strata suitable as a source of water supply. Similar warming effects of bodies of water are also observed in some lakes. In prospecting for water supply the ground near large rivers and lakes, therefore, should be thoroughly explored.

Shallow and sluggish streams usually freeze earlier than the deeper and more rapidly flowing rivers. In the Arctic and Subarctic regions ice may form not only on the surface of the river but at the bottom of the channel as well. This is called bottom ice or anchor-ice. Bottom ice may form even before the surface ice begins to appear. The physics of its formation is not yet fully understood.

Locally where a large spring issues in a river channel the river may remain unfrozen through a part or all the winter. In searching for water supply such unfrozen windows in the river ice should receive careful study as they usually indicate the site of subaqueous springs which, as a rule, afford dependable sources of water throughout the year.

As the winter freezing of a river sets in, the live channel beneath the ice gradually becomes constricted and eventually becomes too small for the entire volume of flowing water to pass through it. The impeded flow creates hydrostatic pressure, which forces the water to permeate the porous alluvium on both sides of the stream. The gradually increasing hydrostatic pressure ultimately causes the water to force its way to the surface either through the river ice or through the ground near the stream or some distance away from the river channel.

Water issuing from such a break spreads out and freezes in successive sheets of ice, which may cover an area of several square kilometers and commonly attain a total thickness of 1 to 4 meters. These ice incrustations or "ice-fields" are known as icings or aufeis, in Russian "naledee" (singular "naled'"). (Fig. 15). The icing that forms over the river ice is generally referred to as river icing, whereas the icing formed on the surface of the ground is known as ground icing. (Fig. 16).

Commonly a more or less prominent mound is formed at the surface outlet of icing water. On a river this mound may be formed partly by the bulge in the river ice. These mounds are known as icing-mounds and are distinguished from the ordinary frost-mounds by the fact that they are composed entirely of ice and only rarely contain soil or rocky material. In some cases the distinction between a frost-mound and an icing-mound, however, may not be very sharp.

Icings may be seasonal or may last a number of years. Such large and persistent icings the natives of Siberia call taryn.

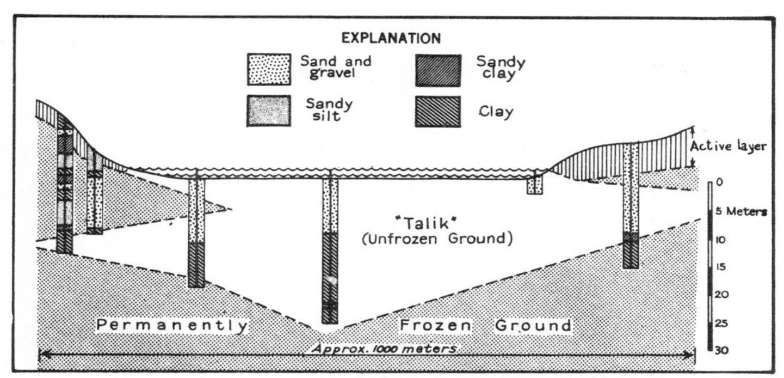

Cross-section through the Gorodskaya protoka (channel) at Yakutsk showing extent of frozen ground

(After Svetozarov)

FIG. 13

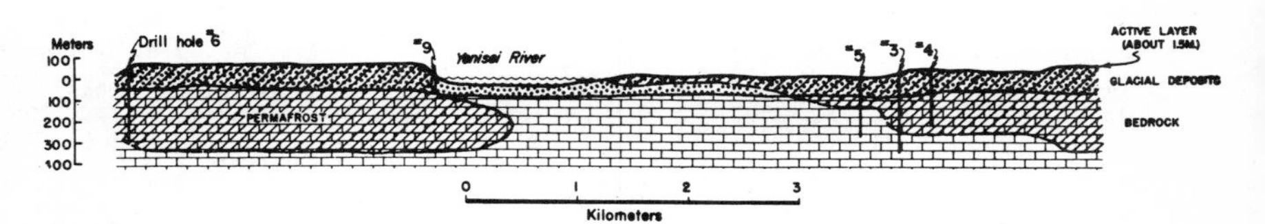

Permafrost profile at Ust'-Port, Siberia.

(After Saks)

FIG. 14

River icing on Tekana River, Siberia.

(From L'vov)

FIG. 15

Icing formed by a spring.

(From L'vov)

Large icings present a formidable obstacle to the travel but, on the other hand, icings from which water flows, or may be induced to flow, throughout the winter may be utilized or developed as dependable sources of water supply.

Lakes are numerous in permafrost regions but the part they play in the hydrologic regime of frozen ground is relatively small. Large lakes have a warming effect on the surrounding ground which thereby remains unfrozen during the winter. Most small lakes usually freeze clear through and ground-ice persists in the bottoms of many of them, even during the summer. Lakes frequently form as a result of melting of ground-ice.

The role of springs in the hydrology of frozen ground is variable. Springs that depend essentially on precipitation flow after a rainy season. Springs fed by the melting of ice and snow flow as long as this source lasts, at times through the summer and well into the winter. Intermittent springs appear where the freezing of superficial ground exerts pressure on the water-saturated layers below and forces the water to the surface. Somewhat similar springs occur near rivers as has already been mentioned in the discussion of icings. Such springs usually appear during the winter and their persistence depends upon local factors.

Springs that are fed by water from within or from below the permafrost usually flow throughout the year and, if not too strongly mineralized, may serve as excellent sources of water supply.

Ground-water in the permafrost regions may be considered under the following headings:

1. above the permafrost (suprapermafrost water)
2. water within the permafrost (intrapermafrost water)
3. water below the permafrost (subpermafrost water)

1. Water above the permafrost extends down to the impervious permafrost table. Its source is primarily rain or melt water, surface water, vapor that condenses near the cold surface of the ground, and finally, seepages from within and from beneath the permafrost. Most of this water freezes and melts with the seasons. In areas where the permafrost table lies at a greater depth than the base of the winter freezing some of this water will remain liquid all through the year.

The behavior of water above the permafrost is different from that of the ground-water in the areas outside of the permafrost regions. During the summer this water moves only by gravity and percolates down the slopes of the irregular surface of the frost table. In winter as ground freezes from the top downward, the water which percolates through the still unfrozen part of the active layer becomes gradually constricted in its channel and becomes subject to a hydrostatic pressure. Thus during the winter the water above the permafrost behaves more like "confined ground-water". The magnitude of this pressure is proportional to the thickness of the overlying frozen part of the active layer. Under this pressure the water may be forced to the surface and form seepages and sheets of ice (icings) or it may be wedged in between the frozen and unfrozen layers where it freezes to form layers or lenses of ground-ice. Occasionally the water above the permafrost may even split the already frozen layers of ground and fill the space between them.

The freezing of the water above the permafrost is the main cause of swelling ground, of frost-mounds, and of icings. The regime of this water should be investigated thoroughly before any engineering construction is undertaken. This water should not be counted on as a possible source of water supply as it usually ceased to flow in the winter and the small amount that may be obtained during the summer is likely to be polluted.

2. The water within the permafrost percolates through the unfrozen layers or vein-like passages as illustrated in the diagram, Fig. 12.

The water within the permafrost is fed either by the infiltration of surface water or by water from beneath the permafrost, or both. It commonly exists in considerable volume and may well yield a steady supply of water. Strongly mineralized water (as for example sea water) may permeate certain layers of permafrost and remain fluid even though the surrounding ground may have a temperature considerably below 0° C. Such occurrences of saline ground-water have been reported from the Arctic coas

In mining operations and other excavations care should be exercised not to pierce permafrost that may contain water.

The water within permafrost may or may not be under hydrostatic pressure.

3. Water beneath the permafrost is always fluid and being overlain by the impervious permafrost it is, as a rule, under considerable hydrostatic pressure. The water below the permafrost is commonly found in alluvial fans at the bottom of recent and old valleys, and, locally, even in bedrock.

Except where it is strongly mineralized the water below the permafrost may serve as a good source of water supply. The main drawback in the exploitation of this water lies in the difficulty of penetrating (drilling) the overlying permafrost, especially where it is more than 100 meters thick.

Ice in Frozen Ground

Permanently or temporarily frozen ground commonly contains ice either in uniformly disseminated minute grains or crystals or as somewhat larger separate inclusions regularly distributed between the layer of ground. Ice in the ground may also occur in large solid masses (ground-ice), usually of sheet-like or lens-like form, or as veins, dike-like wedges, and pipes. Figs. 17 and 18.

Ground-ice intercalated with frozen silt, Fairbanks District, Alaska.

(From the U.S.Geol.Survey files)
FIG. 17 (Photo by Taber)

Lenses, veins, and wedges of ground-ice in the Yukon region, Alaska.

(From the U.S.Geol. Survey files)
FIG. 18 (Photo by Capps)

From the standpoint of engineering, the content of ice in frozen ground is of paramount importance; the thawing of this ice produces excessive wetting and undesired plasticity of the ground and renders it unstable and susceptible to settling, caving, and even flowing. It is therefore important that all necessary data regarding the ice- or water-saturation point of the ground should be ascertained before any construction is undertaken.

According to Bouyoucos the fundamental principles that govern the formation of ice in the ground are:

1. Tendency of free water particles in the pores of the ground to freeze at a normal temperature of 0° C. or even slightly lower, from -1° to -4° C.
2. Tendency of a part of the water contained in fine capillaries to resist freezing at exceedingly low temperatures, down to as low as -78° C.
3. Tendency of water in the process of freezing to pull minute particles of water from adjacent capillaries in which they do not freeze at normal temperatures.

The freezing temperature of water is also controlled by the amount of dissolved salts. For example, the increase of NaCl in solution will lower the freezing point of water as shown in the following table and graph (Fig. 19)

Amount of NaCl in grams per 100 cm³ of water.		0.0000	0.1208	1.479	3.5*	6.5	6.7	*10.5*	10.7	*16.8*	*21.*	22.9	30.4
Freezing temperature	Centigrade	-0.0000°	-0.0736°	-0.8615°	-1.9°	-3.3°	-3.6°	*-6.67°*	-6.32°	*-12.2°*	*-17.8°*	-14.77°	-21.12°
	Fahrenheit	32°	31.87°	30.45°	28.5°	26.06°	25.52°	*20.0°*	20.6°	*10.0°*	*0°*	5.42°	6.0°

*/ Normal salinity of sea water
Vertical figures based on Russian sources
Slanting figures taken from *Jessup, Refrig. Eng., 12, 171 (1925)*

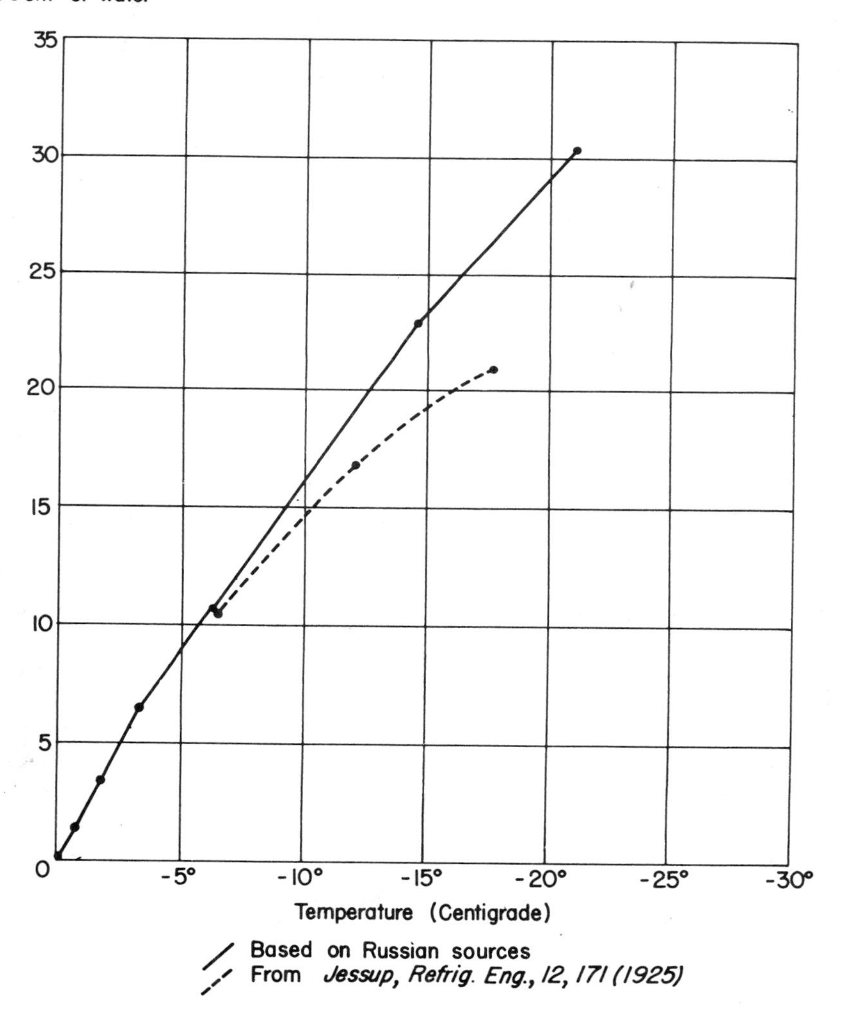

Freezing points of NaCl solutions

FIG. 19

The amount of ice that can form in the ground through the freezing of water depends on several factors of which the most important are:

1. Amount of interstitial water in the ground before freezing.
2. Available supply of water in the adjacent ground.
3. Texture of the ground.
4. Porosity of the ground.

As would be expected, there appears to be a definite relationship between the abundance of ground-ice and the conditions that are favorable for the accumulation of ground-water. In low places and near the bottom of valleys ice lenses and layers are more common than in well drained areas along the slopes of mountains.

Formation of thick layers or thin intercalations of ice in the groun may take place under a variety of conditions of which the most common are

1. Formation of ice by freezing of surficial water after its penetration into the ground.
2. Formation of ice by freezing the ground-water that is brought up by seepage from deeper levels.
3. Formation of ice by freezing of surface waters such as lakes or rivers and possibly also of bottom-ice of rivers and lakes that are subsequently covered by sediments or by landslides.
4. Burying beneath sediments ice that has already formed by any means.
5. Through burial of glacial ice, of snow-banks, or icings or even of sea ice by later sediments or by landslides.

Thus, ice can become imbedded in the ground not only as a result of certain hydrographic and hydrologic processes but through surficial ground movements as well.

Ice usually possesses a distinctive structure that is characteristic of the environment in which it formed.

1. Compact crystalline ice is formed by quiet freezing in large basins and is most commonly found in the middle of the ice layer
2. Acicular ice, commonly formed at the bottom of an ice layer (near the contact with water), consists of numerous long crystals and hollow tubes of various forms, occasionally showing layered arrangement and containing bubbles of air. Due to its silky appearance this type of ice is also known as fibrous ice or satin ice.
3. Layered ice is produced by compaction of separate layers of wet snow or by periodic freezing of separate layers of water.
4. Firn ice, consisting of separate lustreless spherical granules measuring several millimeters in diameter, is produced by the freezing of snow.
5. Fine aggregate structure in ice is usually formed by freezing of stirred water. It is commonly found in the upper layer of ice formed in a large basin.
6. Loose-flaky ice is found in fresh snow and is also formed by freezing of water that is condensed from vapor.

One should be forewarned, however, that it is not always possible to recognize the origin of a particular mass of ice by its structure because the ice may undergo a change under pressure and its original structure may be either obscured or completely obliterated.

Permafrost and vegetation

In addition to the general climatic and geographic factors which control the distribution of the different types of vegetation, the areal distribution of plants in the arctic and subarctic regions is also influenced by the factors which are directly connected with the presence and relative extent of permafrost and the duration of seasonal frost in the ground.

The permafrost has a marked effect on the features of the peat bogs which are so widespread in the arctic and subarctic regions. The rate of growth of sphangnum decreases as one goes further north. Peat bogs are, as a rule, free of trees as they cannot endure the excessive moisture and only the drier parts of the bogs support dwarfed and stunted trees.

The lichen-peat bogs are widespread north of the 64th parallel which is the treeless area of shallow permafrost with very low soil temperatures.

There appears to be no marked difference in the grassy vegetation between the areas underlain by permafrost and those in which permafrost is absent.

Trees, like pine, with prominent tap roots will not thrive where permafrost is present. Pine groves can be generally taken as a good indication that the permafrost is either absent or is present at a considerable depth below the surface. Larch or tamarack also has a tap root but it can exist in areas where permafrost is only a short distance below the surface. The reason for the survival of larch in the permafrost areas lies in its ability to form auxiliary roots on the newly buried part of the stem. The rising level of the permafrost kills the pine but in the larch the destroyed roots are replaced by a new system at a higher elevation.

Larch frequently grows in association with spruce which is characterized by a very shallow system of roots. For this reason spruce may thrive in areas where permafrost is only a foot or 18 inches below the surface.

The presence or absence of permafrost may also be suggested by the following criteria.

Fir trees commonly grow where permafrost is absent or where it is present at a considerable depth below the surface of the ground.

Dwarfed and stunted birches usually indicate presence of permafrost close to the surface.

Willow groves generally point to the absence of permafrost and to the presence of ground-water which freezes only for a short time.

"Drunken forest" (Fig. 29) - a group of irregularly inclined trees - normally indicates the presence of frost-mounds or strongly swelling ground.

Trees growing on a slope may have curved trunks indicating the creep (or solifluction) of the ground along the permafrost table. Orientation and areal distribution of curved trunks should be noted and the age of straight and curved trees should be determined.

Trees on pingo (large frost-mound) may enable one to determine how old the mound is. In the permafrost province the depth of tree roots approximately corresponds to the thickness of the active layer making it possible to get some idea of the thickness of the water bearing zone above the permafrost.

Peat and moss usually indicate a relatively thin water-bearing zone above permafrost.

In steppe areas various other plants may be used as indicators of the presence of ground-water close to the surface.

Physical and Mechanical Properties of Frozen Ground

Physical properties of frozen ground depend on its composition, texture, content of ice, and temperature, but there is as yet no classification in which all these factors are taken into account. Furthermore, in engineering practice it is equally important to consider the properties of unfrozen or thawed ground. For example, a building erected on permafrost may cause thawing and therefore should be so designed that it will remain undamaged under the changed condition of the ground.

Moreover, thawed ground is slightly different in its physical properties from ground that has not been subjected to freezing even though they may be alike in composition, texture, and moisture content. This difference is, apparently, due to the fact that some time elapses before thawed ground "settles" or attains its normal structure on which depends its ability to support a load. Furthermore, the presence of the underlying still frozen ground makes a difference in the physico-mechanical properties of the unfrozen ground directly above.

It is an established fact that the compressive strength of the ground frozen for the second time is considerably less than that of ground that has been frozen only once. Solid bedrock is comparatively little affected by freezing and is therefore, for the most part, left out of the consideration. The following brief discussion is devoted almost entirely to normally unconsolidated materials listed on page 35.

CLASSIFICATION OF CLASTIC AGGREGATES

(Granulometric classification of grounds)
(Based on Russian sources)

NAME	% of grain-sizes			REMARKS	
	< .005 mm	.005 - .05 mm	.05 - 2 mm		
"Fat" clay	> 60	-	< 3	Physical properties wholly dependent on amount of ice	When thawed are little permeable, settle (compact) very slowly, several years or decades - if oversaturated upon melting will ooze out from beneath the foundation.
Clay	60 - 30	Less than half of remaining 40 - 70%	More than half of remaining 40 - 70%	Physical properties wholly dependent on amount of ice	
Silty clay	> 30	More than each of the other two sizes	-	When saturated or oversaturated with water turns to slud (=sludge)	
Silty-sandy clay	30 - 10	More than half of remaining 70 - 90%	Less than half of remaining 70 - 90%	When saturated or oversaturated with water turns to slud (=sludge)	
Sandy clay	30 - 10	Less than half of remaining 70 - 90%	More than half of remaining 70 - 90%		
Silt	< 3	> 50	-	When saturated or oversaturated with water turns to slud (=sludge)	
Clayey sand	10 - 3	-	> 50	Permeable; settles quickly, time element not important	
Silty-clayey sand	10 - 3	More than half of remaining 90 - 97%	Less than half of remaining 90 - 97%	When saturated or oversaturated with water turns to slud (=sludge)	
Silty sand	< 3	20 - 50	-	When saturated or oversaturated with water turns to slud (=sludge)	
Sand	< 3	< 20	-	Permeable; settles very little if at all	

Grounds containing 10 or more percent of gravel are designated as gravelly.

Ice fills some or all of the interstitial space between the mineral grains in frozen ground and acts as a cement. Mechanical properties of frozen ground, therefore, tend to approach those of ice.

The relative strength of ice under different stresses depends on its structure, temperature and the nature of surrounding conditions. Compressive strength of ice also depends on the rate of increase of pressure and on the direction of forces with reference to the axes of ice crystals. Ice will withstand a much greater pressure if the force is directed parallel to the long axis of a crystal that when it is perpendicular to this axis. In ice formed over a body of water the long axes of crystals are normal to the water surface. For example, river ice which has been tested by Pinegin gave the following results:

ULTIMATE COMPRESSIVE STRENGTH OF ICE in kg/cm^2

	Temperature in C°	Force applied parallel to the long axis of a crystal	Force applied normal to the long axis of a crystal
Top of ice layer	0° to -2° -8° to -10° -20° to -23°	20.7 28.0 38.4	18.4 25.2 28.2
Middle of ice layer	0° to -2° -8° to -10° -20° to -23°	35.8 32.8(?) 76.0	28.2 33.5 69.2
Bottom of ice layer	0° to -2° -8° to -10° -20° to -23°	17.5 20.4 37.6	12.0 18.2 32.1

The extreme range of variation in the strength of ice depending on its structure was demonstrated by Finlayson. River ice with well developed structure, at -1.6° C. had a compressive strength of 127 kg/cm^2 when a force was applied parallel to the crystal axis and only 74 kg/cm^2 when the force was directed at right angle to the axis.

According to Tsytovich and Sumgin the rate of increase of force has the following effect on the strength of ice:

With the increase of load of 20 kg/cm^2/min. the compressive strength of ice was 60 kg/cm^2

" 36 " .... 37 "

" 50 " .... 24 "

The tensile strength of ice is also dependent on structure, temperature, and the orientation of the crystals.

Another property of ice, important in some engineering problems, is the bending strength of ice. Experimental data on this property is shown in the following table:

Limits of Bending Strength of Ice in kg/cm^2 (From Tsytovich and Sumgin)						
Temperature of ice in C^o	-18.7	-9 to -11	-3 to -5	-4	-0.2	about 0
Load under which bent ice breaks in kg/cm^2	32 to 34	33	18	34	16	10 to 14

The bending strength of ice depends more directly on the orientation of crystals than is the case with other types of stresses in ice. Two pieces of the same river ice with a change of orientation of 90° varied in their bending strength from 4 to 29.6 kg/cm^2.

In general, the amount of deformation in ice is directly proportional to the load and inversely proportional to the coefficient of viscosity. As can be seen from the diagram (Fig. 20), the coefficient of viscosity of ice varies markedly with temperature.

Frozen ground with pores completely filled with ice is susceptible to deformation, the amount of which varies with temperature and pressure. With the rise of temperature the deformation of ice becomes more intense and its plasticity and flowage become more pronounced. When the ice begins to melt the character of the ground is completely changed and its deformation is greatly intensified even without application of additional load. The rise of temperature to the melting point may thus have a stronger effect on the stability of frozen ground than the increased load.

The compressive strength of frozen ground increases with the lowering of temperature. Compressibility of sand varies with the amount of moisture (ice) and reaches a maximum when the pores are completely filled with ice. The compressive strength of clays decreases with the increase of moisture content.

The tables on the following pages illustrate the behavior of different frozen grounds under compression.

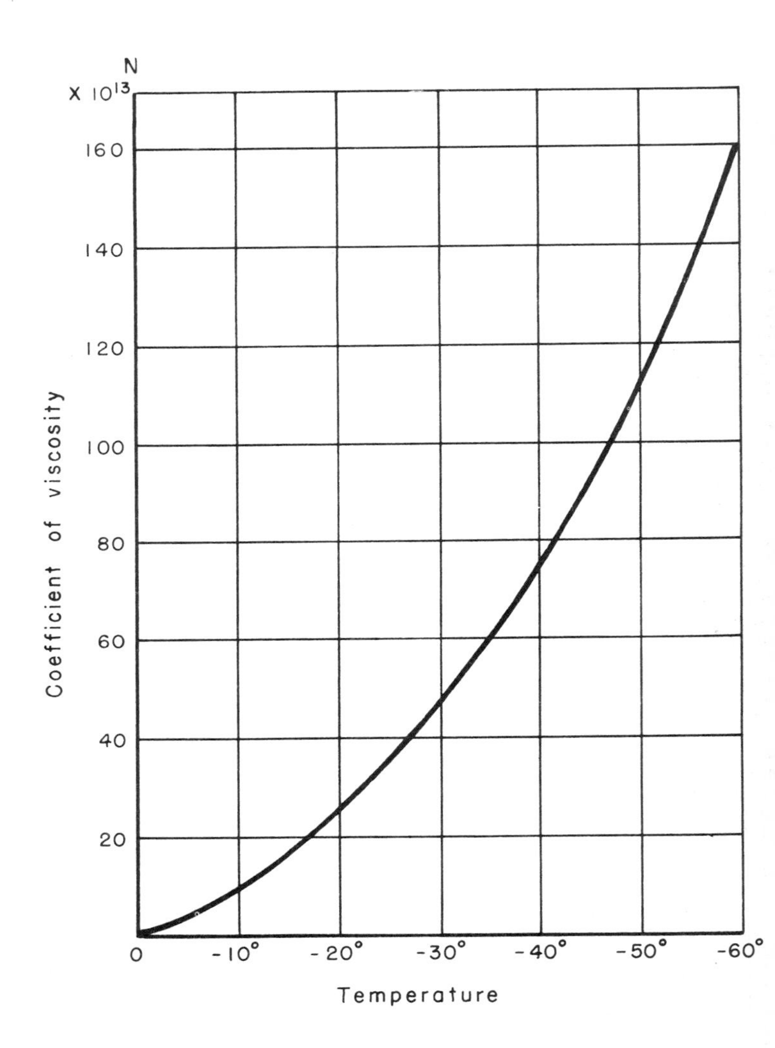

Coefficient of viscosity of ice in relation to temperature

(From Tsytovich and Sumgin)

FIG. 20

Compressive Strength of Frozen Grounds at Temperatures from -0.3° to -2° C.

(From Tsytovich and Sumgin)

Type of Ground	Temperature in C°	Moisture (ice) in %	Ultimate Compressive strength in kg/cm²
Sand with rubble (Grains:>1mm=85%)	-1.4	15	27
Sand (Grains:>1mm=80%)	-0.8	18	28
Sand, arkosic (Grains:>1mm=61%)	-1.6	15	43
Sand, arkosic (Grains:>1mm=44%)	-1.7	20	30
Silty sand (Grains:<0.005mm=4%)	-0.4	38	22
Clayey sand (Grains:<0.005mm=10%)	-1.2	46	12
Clayey sand (Grains:<0.005mm=10%; >1mm=25%)	-0.4	17	10
Clayey sand (Grains:<0.005mm=10%; >1mm=5%)	-0.9	31	26
Silty, clayey sand (Grains:<0.005mm=7%; >1mm=0.5%)	-1.1	20	28
Clayey sand (Grains: 1-0.05mm=68%; <0.005mm=8%)	-0.3	21	12
Clayey sand (Grains: 1-0.05mm=68%; <0.005mm=8%)	1-1.1	21	25
Clayey sand (Grains:<0.005mm=10%)	-0.7	36	15
Clayey sand - -	-1.1	32	18
Clayey sand (Grains:<0.005mm=9%; >1mm=15%)	-1.5	48	23
Sandy, micaceous clay (Grains:<0.005mm=13%; >1mm=9%)	-0.6	32	19
Sandy clay (Grains:<0.005mm=14%)	-0.8	32	24
Sandy clay with rubble (Grains<0.005mm=17%; >1mm=12%)	-1.8	23	22
Sandy clay (Grains:<0.005mm=18%; >1mm=14%)	-1.9	27	26
Sandy clay (Grains:<0.005mm=15%; >1mm=19%)	-1.3	21	20
Sandy clay (Grains:<0.005mm=20%; >1mm=21%)	-0.6	37	22
Sandy clay (Grains:<0.005mm=22%)	-1.4	43	22
Sandy clay (Grains:<0.005mm=23%)	-1.9	41	27
Sandy clay (Grains:<0.005mm=14%)	-0.8	28	17
Sandy clay (Grains:<0.005mm=15%)	-1.0	34	15
Sandy, silty clay (Grains:<0.005mm=26%)	-2.0	52	29
Sandy clay (Grains:>0.005mm=17%; >1mm=16%)	-0.8	26	23
Sandy clay (Grains:>0.005mm=15%; >1mm=3%)	-0.8	37	18
Sandy clay (Grains:<0.005mm=17%; >1mm=17%)	-1.6	29	24
Sandy clay (Grains:<0.005mm=17.5%)	-1.3	24	17
Silt (Grains: 0.05-0.005mm=63%;<0.005mm=14%; organic matter = 18%)	-0.3	59	5
Silt (Grains: 0.05-0.005mm=63%;<0.005mm=14%; organic matter = 18%)	-1.1	52	12
Silt (Grains:<0.005mm=12%)	-0.7	40	12
Silt (Grains:<0.005mm=14%)	-1.2	38	14
Silt (Grains:<0.005mm=22%)	-1.6	52	24
Silt (Grains:<0.005mm=13%)	-1.0	70	22
Clay (Grains:<0.005mm=36%)	-0.3	43	6
Clay (Grains:<0.005mm=36%)	-1.5	48	16
Clay (Grains:<0.005mm=30%)	-1.7	34	20
Clay (Grains:<0.005mm=26%)	-1.5	42	17
Clay (Grains:<0.005mm=45%)	-1.7	45	15

As can be seen from the preceding table, the compressive strength of water (ice)-saturated frozen grounds at the temperature not exceeding -2° C. is relatively small, varying from 5 to 30 kg/cm^2.

The values given in the table may be summarized as follows:

Mean Values of Compressive Strength of Water-Saturated Frozen Grounds

Ultimate Compressive strength in kg/cm^2			
Type of Ground	At temperatures not lower than -0.5° C.	At temperatures from -0.5° to -1.5° C.	At temperatures from -1.5° to -2.0° C.
Sands	22	27	30
Clayey sands	11	22	-
Sandy clays	-	20	26
Clays	6	17	-
Silts	5	15	23

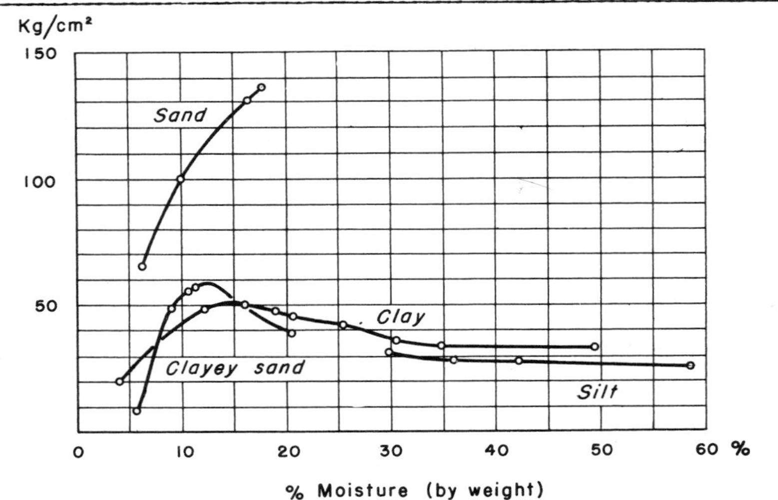

Compressive strength of frozen grounds in relation to amount of moisture.

(After Tsytovich and Sumgin)

FIG. 21

Elastic and Plastic Deformation of Frozen Grounds under Compression

Type of ground	Moisture (ice) content in %	Temp. in C°.	Limits of load in $\frac{kg}{cm^2}$	Time in min. from beginning of application of load	Relative deformation		Elastic deformation in % of total deformation
					Elastic	Plastic	
rozen clay	36.8	-2.8	0.5-1.5	6	0.000066	0.000074	49
				7	0.000061	0.000129	32
		-2.8	0.5-2.5	10.5	0.000127	0.000719	15
		-2.8	0.5-2.5	11.5	0.000142	0.000562	20
		-2.4	0.5-3.5	12	0.000220	0.001350	14
			0.5-3.5	13.5	0.000200	0.001250	14
	32	-2.0	0.5-1.5	11.5	0.000110	0.000190	37
		-1.9	0.5-2.5	12.5	0.000170	0.000845	17
		-1.5	0.5-3.5	13.5	0.000220	0.002050	10
		-1.4	0.5-3.5	14.5	0.000200	0.003560	5
	32.5	-1.3	0.5-2.5	30	0.000055	0.006950	1
ozen ayey sand	17	-12.2	0.5-1.5	11	0.000010	0.000010	50
		-10.4	0.5-2.5	12	0.000030	0.000020	60
		-8.7	0.5-3.5	13	0.000055	0.000070	44
		-7.0	0.5-4.5	14	0.000115	0.000205	36
	12.8	-1.7	0.5-2.5	17	0.000042	0.000076	33
		-1.1	0.5-3.5	18	0.000111	0.000231	32
	16.9	-2.5	0.5-2.5	30	0.000180	0.000475	27
		-1.7	0.5-3.5	30	0.000350	0.006670	5
ozen silt	39.2	-4.9	0.5-1.5	11	0.000035	0.000001	97
		-3.5	0.5-2.5	12	0.000090	0.000045	67
		-3.4	0.5-3.5	17	0.000190	0.000140	58
			0.5-5.0	34.5	0.000430	0.000730	37
	38.9%	-1.0	0.5-3.5	18	0.000335	0.000260	56
		-0.8	0.5-5.0	19.5	0.000640	0.003030	17

(From Tsytovich and Sumgin)

The shearing strength of frozen grounds depends primarily on temperature and to a much less extent on the content of moisture (ice) and the texture of the ground.

The increase of moisture (ice) content tends to increase the shearing strength of a ground but after a certain maximum further addition of water (ice) tends to reduce the shearing strength of the material.

See tables on the following pages.

Shearing Strength of Ice-saturated Frozen Grounds

(From Tsytovich and Sumgin)

Name and composition of ground	Temperature in C°	Moisture (ice) content (by weight) in %	Ultimate Shearing strength in kg/cm²
1. Clay (Grains<0.005 mm=45%)	-2.1	35	6.6
2. Clay (Grains<0.005 mm=41%)	-1.8	33	8.0
3. Clay (Grains>1 mm=6%;<0.005 mm=31%)	-1.9	29	9.0
4. Sandy clay (Grains<0.005 mm=27%)	-1.9	28	8.9
5. Sandy-silty clay (Grains<0.05 mm=25%)	-2.0	36	8.0
6. Sandy clay (Grains<0.005 mm=22%)	-1.9	34	9.0
7. Sandy clay (Grains<0.005 mm=17%; >1 mm=13%)	-2.1	31	8.5
8. Sandy clay (Grains<0.005 mm=17%; >1 mm=16%)	-1.8	27	8.0
9. Sandy clay (Grains<0.005 mm=15%)	-1.9	23	10.0
10. Sandy clay (Grains<0.005 mm=15%)	-1.6	29	7.0
11. Sandy-silty clay (Grains<0.005 mm=15%)	-1.3	23	6.0
12. Sandy-silty clay (Grains<0.005 mm=15%)	-2.8	23	14.0
13. Sandy clay (Grains<0.005 mm=14%)	-1.7	24	8.0
14. Sandy-silty clay (Grains<0.005 mm=14%)	-1.5	34	7.4
15. Sandy clay / slud/ (Grains<0.005 mm =14%;>1 mm=12%)	-1.7	17	10.3
16. Sandy clay with rubble (Grains>1 mm =25%;<0.005 mm=13%)	-1.6	19	10.8
17. Sandy clay (Grains<0.005 mm=11%)	-1.7	34	8.9
18. Sandy clay (Grains<0.005 mm=10%)	-2.0	39	9.5
19. Silty-clayey sand (Grains<0.005 mm=4%)	-1.6	26	10.0
20. Rubble (weathered granite)(Grains>1 mm =44%)	-1.8	23	11.0
21. Silt (Grains 0.01-0.005 mm= 68%; <0.005 mm=14%)	-0.6	55	7.8
22. Sandy clay (Grains>0.25 mm=13.6%; <0.005 mm=12%)	-0.9	37	8.9
23. Sandy clay with rubble (Grains>0.25 mm = 33%;<0.005 mm=9%)	-1.1	49	15.0
24. Sand (Grains>0.25 mm=51%)	-0.7	18	10.9
25. Sand (Grains>0.25 mm=34%;<0.005 mm=3%)	-0.8	36	12.2

Effect of Temperature on the Shearing Strength of Frozen Grounds

(From Tsytovich and Sumgin)

Name of ground and its granulometric composition	Temperature in C°	Moisture (ice) content in % (by weight)	Ultimate Shearing strength in kg/cm^2
Clay (Grains 0.01 - 0.005 mm=50%; <0.005 mm=36%)	-0.4	45.5	3.7
" " " "	-1.8	50.6	17.2
" " " "	-3.0	49.8	20.9
" " " "	-4.9	44.0	24.3
" " " "	-6.3	42.0	28.5
" " " "	-8.8	45.9	33.5
Clayey sand (Grains 1-0.05 mm=68%; <0.005=8%)	-0.4	18.4	4.9
" " " "	-0.9	17.8	10.6
" " " "	-3.1	19.1	21.8
" " " "	-3.9	16.9	24.8
" " " "	-6.7	19.0	44.2
" " " "	-8.5	16.2	47.5
" " " "	-9.3	19.0	48.5
Clean artificial ice	0.0	-	9.9
" " "	-0.4	-	11.0
" " "	-2.9	-	27.4
" " "	-4.4	-	32.5
" " "	-6.1	-	38.5
" " "	-10.1	-	56.2

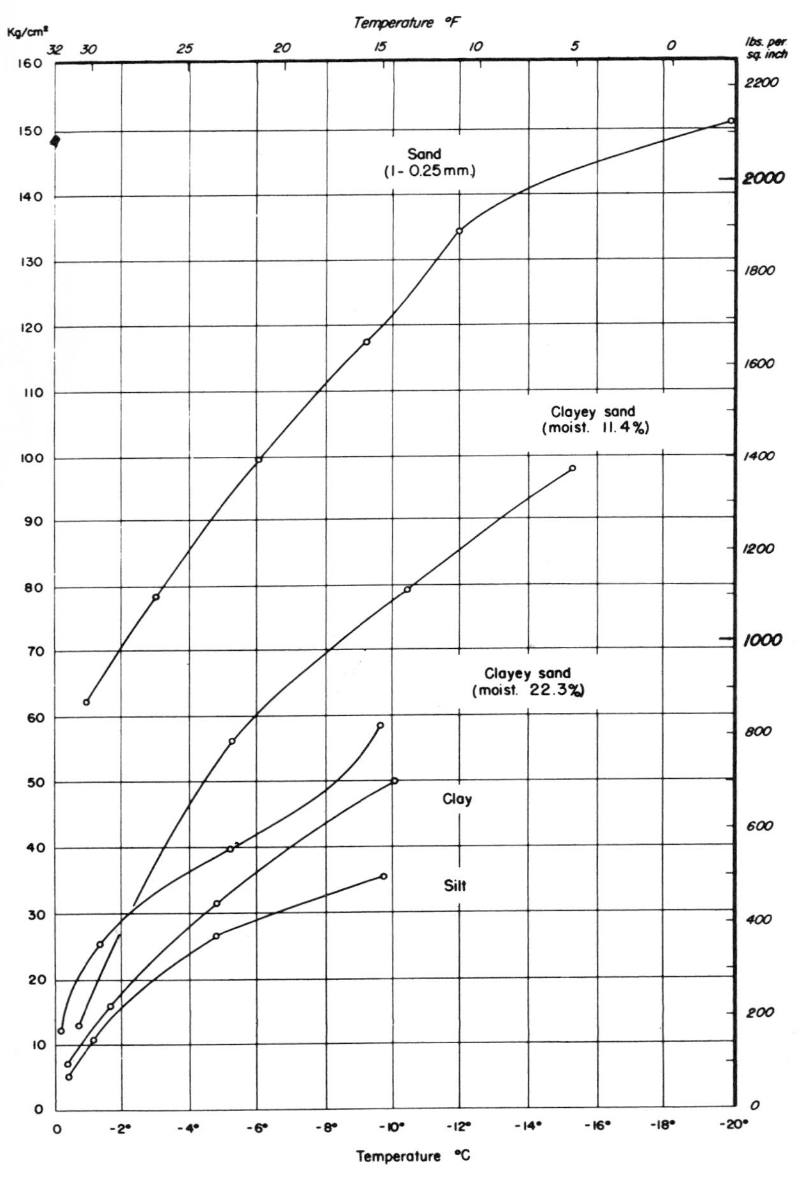

Ultimate Compressive Strength of Frozen Grounds in Relation to Temperature.

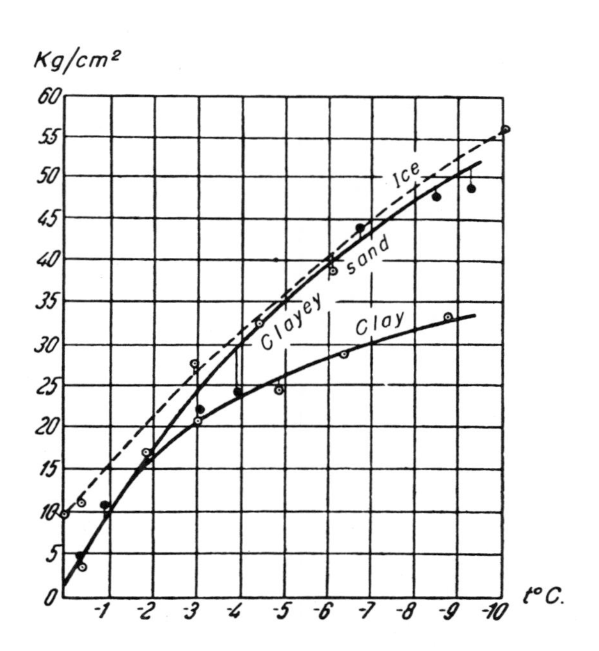

Shearing strength in relation to temperature.

FIG. 23

Adfreezing strength of frozen ground is defined as the resistance to the force that is required to pull apart the frozen ground from the object to which it is frozen. However in engineering practice the tangential adfreezing strength is of greater importance. This is the resistance to the force that is required to shear off an object which is frozen to the ground and to overcome the friction along the plane of its contact with the ground. Care should be exercised in evaluating the tangential adfreezing strength as given in the original Russian reports for in the Russian permafrost literature the tangential adfreezing strength is generally referred to as "sila smerzania" or simply "adfreezing strength".

The tangential adfreezing strength of frozen ground varies with:

1. The amount of moisture in the ground.
2. Temperature of the ground.
3. Texture and porosity of the ground.
4. The nature of the surface of the building material used (smooth or rough).
5. Porosity of the material near the surface.
6. Degree of saturation.

The maximum of adfreezing strength in most grounds is reached at about the maximum saturation of ground with ice. Further increase in the amount of ice, beyond the maximum saturation point, tends to decrease the adfreezing strength, gradually approaching that of pure ice.

The heaving force of ground in the process of freezing is proportional to the adfreezing strength of that ground with the material of foundation. The tangential adfreezing strength varies with the texture of the ground and, in general, is greater in fine and medium-grained sands, than in the coarse-grained aggregates. Adfreezing strength in clays and silts is slightly less than in sands.

The following tables and graphs give the quantitative data on the adfreezing strength between different grounds and wood and concrete under different conditions of temperature and amount of moisture (ice).

Effect of Temperature and Moisture Content on the
Tangential Adfreezing Strength
Between Different Grounds and Water-saturated Wood and Concrete

(From Tsytovich and Sumgin)

Temperature in C°	Type of ground	With water-saturated wood		With water-saturated concrete	
		% of moisture	Adfreezing strength in kg/cm^2	% of moisture	Adfreezing strength in kg/cm^2
-0.2°	Silt (Grains: 0.05-0.005 mm=63%)	29.9	3.6	-	-
	Clay (Grains:<0.005 mm=36%)	27.1	2.9	-	-
	Clayey sand (Grains: 1-0.05 mm=68%)	12.1	1.3		
	Silt (Grains: 0.05-0.005 mm=63%)	22.4	7.0	16.4	4.4
	Silt (Grains: 0.05-0.005 mm=63%)	32.6	8.9	33.0	6.0
	Silt (Grains: 0.05-0.005 mm=63%)	43.8	7.1	44.0	9.2
	Silt (Grains: 0.05-0.005 mm=63%)	51.2	7.6	53.2	3.1
-1.2°	Clay (Grains:<0.005 mm=36%)	22.4	3.2	17.8	7.8
	" " " "	26.4	5.9	26.3	4.8
	" " " "	37.3	13.0	36.2	6.4
	" " " "	56.5	11.8	43.9	5.8
	Clayey sand (Grains: 1-0.05 mm=68%)	6.7	2.8	5.8	2.8
	" " " "	10.1	4.1	11.7	6.4
	" " " "	13.3	7.2	12.1	7.0
	" " " "	16.5	8.2	16.1	11.1
	Silt (Grains: 0.05-0.005 mm=63%;<0.005 mm=14%; organic matter=18%)	18.8	6.9	17.4	7.8
	" " " "	33.9	14.1	32.5	21.8
	" " " "	41.5	28.7	46.4	26.2
	" " " "	51.0	34.8	51.8	28.1
	" " " "	62.2	34.7	58.3	27.7
	Clay (Grains:<0.005 mm=36%)	18.4	12.8	18.9	20.6
-10.0°	" " " "	21.6	15.7	25.1	21.9
	" " " "	28.4	18.6	34.6	25.3
	" " " "	41.4	32.2	46.1	20.1
	" " " "	55.6	31.9	-	-
	Clayey sand (Grains: 1-0.05 mm=68%;<0.005 mm=8%)	5.7	7.9	7.5	10.0
	" " " "	10.1	12.6	11.9	22.8
	" " " "	13.9	21.4	18.1	24.2
	" " " "	19.9	32.3	23.8	21.0
	" " " "	33.5	33.5	-	-

Tangential Adfreezing Strength Between Different Frozen Ground and Water-saturated Wood

(From Tsytovich and Sumgin)

Name of ground	Granulometric composition		Temperature in C°.	% Moisture	Adfreezing strength in kg/cm^2
	% of grains >1 mm	% of grains <0.005 mm			
Clay	None	45	-1.5	41	5
"	13	45	-1.0	39	6
"	-	41	-1.0	30	5
"	6	31	-2.2	29	7
"	2	30	-1.6	24	7.2
Sandy clay	None	27	-0.8	35	4
" "	"	25	-1.2	26	5
Silty, sandy clay	"	24.8	-1.5	40	6
Sandy clay with layers of ice	"	23.5	-0.8	39	4
Sandy clay	"	22	-1.8	39	6
" "	"	22	-1.6	34	4
Sandy, silty clay with layers of ice	"	22	-1.5	43	6
Sandy clay	"	20	-0.5	20	2
Sandy clay with layers of sand	"	18	-1.0	18	5
Silty, sandy clay	14	18	-2.0	25	7
Sandy clay	13	17	-1.1	31	4
" "	17	17	-2.2	25	10
" "	None	16	-0.6	27	4
" "	"	15	-1.6	28	7
" "	"	15	-0.7	25	3
" "	3	15	-0.5	25	2
" "	3	15	-4.0	26	4.3
" "	-	14	-1.2	24	6
" "	None	14	-0.8	36	3
" "	"	14	-2.0	33	5
Sandy clay, micaceous	1	14	-1.8	32	3
" " "	9	13	-0.5	25	2
Sandy clay	25	13	-1.6	17	7
" "	-	11	-0.9	27	5
" "	23	10	-1.6	25	5
Sandy clay, micaceous	None	10	-1.0	39	3.3
Sandy clay	5	10	-1.8	23	7.2
Clayey sand with silt	-	9	-1.1	28	3.1
" " " "	0.5	7	-1.0	17	5.4
Clayey sand	41	7	-1.5	16	1.3
Silty, clayey sand with layers of ice	-	4	-1.6	27	3.3
Granitic arkose	61	?	-1.7	14	2
Gravel	80	3	-1.1	12	3.3

Effect of Temperature on Tangential Adfreezing Strength of Different Materials

(From Tsytovich and Sumgin)

	Temperature in C.	% moisture by weight	Tangential adfreezing strength kg/cm^2
Ice and smooth-surfaced wood (wood was placed in water in air-dry condition)	-1	-	5.0
	-5	-	6.2
	-7	-	11.6
	-10	-	13.7
	-20	-	22.0
Ice and smooth concrete	From -5 to -10	-	9.8
Clay (Grains<0.005 mm=36%) and water-saturated wood (moisture content of grounds about ½ of saturation)	-0.2	27.1	2.9
	-1.5	26.4	5.9
	-5.8	28.4	11.1
	-10.8	28.4	18.6
	-17.8	25.8	29.4
Clayey sand (Grains: 1-0.005 mm=68%;<0.005 mm=8%) and water-saturated wood	-0.2	12.1	1.3
	-1.2	13.0	7.0
	-2.7	10.1	11.0
	-5.2	14.8	19.6
	-5.6	12.9	20.8
	-10.7	14.1	24.7
	-17.4	12.8	27.4
Silt (Grains<0.005 mm=14%; organic matter=18%) and water-saturated wood	-0.2	29.5	3.6
	-0.5	33.4	6.1
	-5.7	34.3	10.6
	-10.3	33.1	14.3
	-12.3	33.2	19.9
	-22.7	34.9	25.9

Type of ground	Dominant grain-size in mm.	Temperature (C°)	Coefficient of saturation in %	Tangential adfreezing strength in kg/cm²
Clay No. 1	0.01	-10°	77	15.3
Sand, fine No. 2	0.25	-10°	76	23.3
Sand, medium No. 3	1.0	-10°	78	26.8
Sand, coarse to fine No. 4	3-0.25	-10°	79	21.7
Sand, coarse No. 5	3-2	-10°	97	19.1
Gravel, fine No. 6	5.0	-10°	77	2.6
Gravel No. 7	10.0	-10°	79	0.9

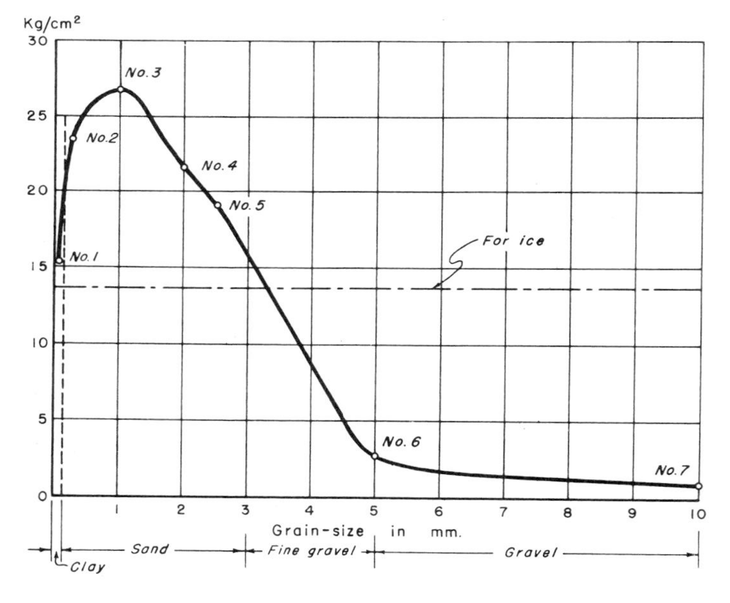

Tangential adfreezing strength between wood and frozen grounds of different texture.

(After Tsytovich and Sumgin)

FIG. 24

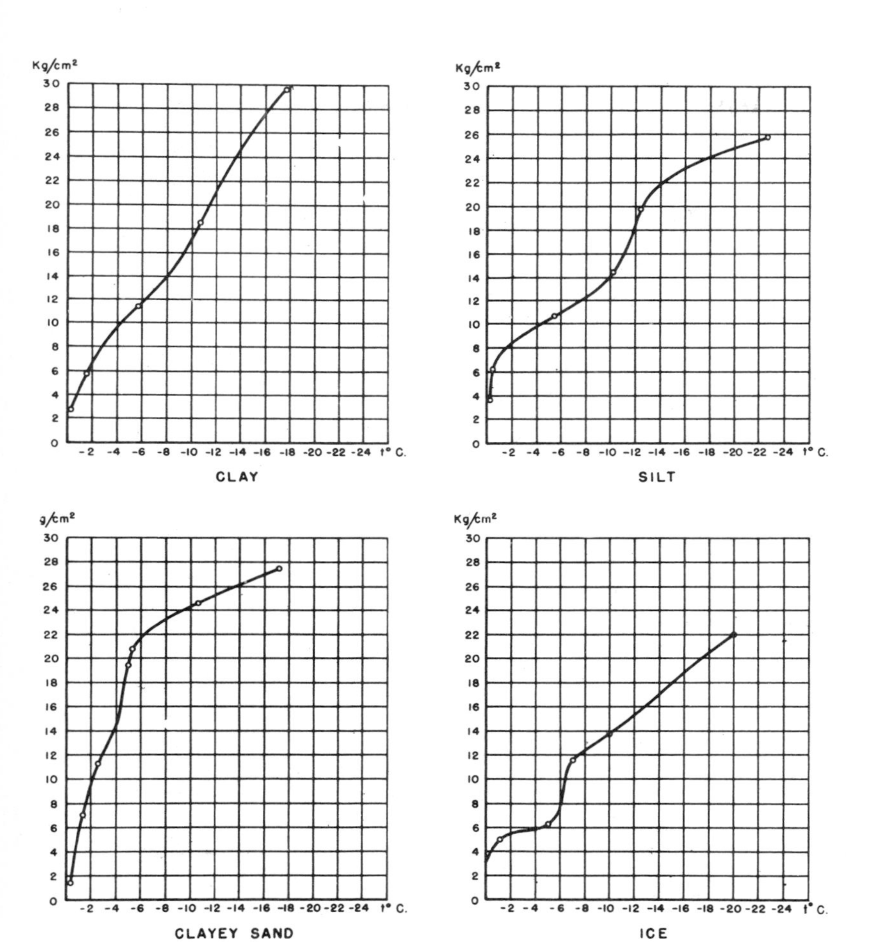

Effect of temperature on the TANGENTIAL ADFREEZING STRENGTH between wet wood and clay, silt, clayey sand, and ice.

(After Tsytovich and Sumgin)

FIG. 25

Values of Permissible Stresses on Ice-saturated Frozen Ground

(From Tsytovich and Sumgin)

Name of ground	Temperature in C°		
	-0.2° to -0.5°	-0.5° to -1.5°	-1.5° tc -2.5°
Sand	3.5 kg/cm²	4.5 kg/cm²	6.0 kg/cm²
Clayey sand	2.0 "	3.5 "	-
Sandy clay	-	3.0 "	4.0 "
Clay	1.5 "	2.5 "	-
Silt	1.0 "	2.0 "	3.0 "

Computational Values of Tangential Adfreezing Strength in kg/cm²

(From Tsytovich and Sumgin)

Adfreezing surfaces	Temperature -1° C.				Temperature -10° C.			
	Ice saturation				Ice saturation			
	0.25	0.50	0.75	1 to 1.4	0.25	0.50	0.75	1 to 1.4
Fine-grained grounds with 8 to 36% of grains of 0.005 mm. (Clayey sand, sandy clay, clay, and silt) with water-saturated wood	2	3	4	6	3	7	13	16
The same grounds with water-saturated concrete	1	2	4	5	7	10	13	13

Heat conductivity of ice is several times greater than that of water. Frozen ground, therefore, has a much higher heat conductivity than the same ground in an unfrozen state.

The coefficient of heat conductivity of ground is the quantity of heat in kilogram calories which pass during one hour through a surface of one square meter of material to another surface of the same material one meter away with one degree difference in temperature between the two surfaces. It should be noted that the data on heat conductivity of ground at different moisture content and different temperature, especially at the temperature below the freezing point, is very scanty. The heat conductivity of a given ground at different temperatures is variable. In the frozen ground at temperatures below zero the heat conductivity gradually rises and at zero degree there is a sharp drop and with the further rise of the temperature the heat conductivity again gradually increases. Similar relationship is also observed in water.

The results of investigations by Professor Evdokimov-Rokotovsky on the heat conductivity of the most characreristic grounds, clay, sand, and of water are shown graphically on fig. 26. On the curves for each ground the moisture content is indicated by weight. From these graphs it can be seen that the heat conductivity at the freezing point on the average increases 3.7 times for water, 3 times for sand and 1.5 times for clay. It will be also observed that the heat conductivity of sand is less than that of clay, approximately ½ of that of clay. Very little information is available on the heat conductivity of peat, which is widespread in the Arctic regions. However it is known that peat can absorb water up to 300% of its volume. It is therefore permissible to assume that the heat conductivity of saturated peat will be close to that of water, which, in the range of temperatures from 0° to 20° Centigrade will be about .5, but in a frozen state it will be about 2.0. Dry peat, which is commonly used as an insulating material, has the heat conductivity coefficient of .06. It will be thus seen that the heat conductivity of peat will vary between wide limits depending upon the amount of moisture, compacting, and whether or not the material is frozen. It can be thus concluded that during summer a greater amount of heat is transferred through the water-saturated peat into the underlying layers of soil than through dry peat. In winter when water-saturated peat is frozen the heat is transferred in the opposite direction, from the soil into the air and the amount of heat passed into the air will thus be four times as great as in the summer. When dried, this negative factor of loss of heat may be considerably decreased. From the above it can be easily seen that such an exchange of heat through peat will have a marked effect on the accumulation of cold in the ground and on the formation of permafrost.

The heat conductivity of ground also depends on its structure. In grounds of similar texture with a decrease of specific gravity, there is a decrease of heat conductivity, thus if it is necessary to decrease the heat conductivity of certain fills of sand or slag in order to warm up the underlying ground the material of the fill should be graded and laid down in layers of even texture. On the other hand to prevent the convectional flow of cold air the successive layers of fill should be as follows;

coarse grained at the bottom, fine grained at the top. In general, sand is used to allow a greater amount of heat to pass into the ground during the summer and to minimize the radiation of heat from the ground during the winter. Sand can be readily drained or can be easily saturated. By wetting the sand during the summer and by draining it during the autumn prior to the frosts, it is possible to appreciably change the thermal regime of the underlying ground.

Heat conductivity of different materials that may be used in various engineering projects in the permafrost area is given on the following page

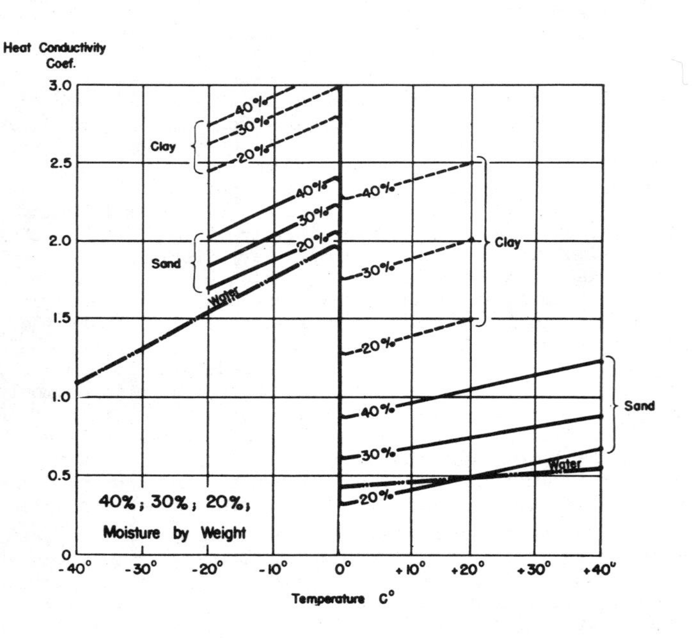

Heat conductivity of CLAY, SAND, and WATER at different temperatures.

(After Evdokimov - Rokotovsk

FIG. 26

Heat Conductivity of Materials

(From Bykov and Kapterev)

Materials	Volumetric weight	Moisture % of volume	Mean temperature in C.°	Coeff. of heat cond. cal/m/h/°C
Concrete (Portland cement, sand, and gravel (1:2:2)	2.18	Dry (dried ½-2 years)	20-30	0.65-0.66
Concrete	1.6	Dry	0	0.72
Concrete (1:5 mixture)	1.9	8.6	0	1.17
Concrete	2.27	10.2	0	1.10
Concrete (mixture 1:12)	2.25	Dried 2 weeks	20-40	0.76
Concrete tamped (cement, sand, stone aggr.)	2.00	Normal	0	1.10
Concrete with brick aggregates	1.90	Normal	0	1.00
Lightweight concrete (1 part cement to 9 parts of porous slag)	0.55	-	20-90	0.19
Reinforced concrete	2.20	Normal	0	1.33
Brick wall (1 brick thick), plastering 28 cm on each side, right after construction	1.96	25	10	1.20
Same after 4.5 months of drying in air	1.76	3.4	10	0.84
" " 6.5 " " " "	1.75	1.9	10	0.74
" " 9 " " " "	1.74	1.0	10	0.64
" " 12.5 " " " "	1.72	0.5	10	0.60
Brick wall - new	1.57-1.63	-	20-40	0.82-0.45
" " - dry	1.42-1.46	-	20-40	0.82-0.45
Old brick masonry	1.85	-	20-40	0.35
Stone masonry	-	-	20-40	1.3-2.1
Masonry with stones of sp.gr. 1.6+	1.20	-	0	0.71
Masonry of light stones	1.00	-	0	0.60
Masonry of granite, basalt, marble	2.50	-	0	2.5
Masonry of limestone, sandstone, etc.	2.20	-	0	1.2
Masonry of light rocks	1.60	-	0	0.71
Masonry of light rocks	1.20	-	0	0.60
Water	1.00	-	78-72.4	0.47-0.58
"	1.00	-	20-40	0.52
"	1.00	-	0	0.50
Ice	-	-	-	2.05
"	0.88-0.92	-	20-40	1.90
"	0.90	-	0	2.00
"	-	-	-	1.50
"	0.92	-	-	1.80-1.86
Air	-	-	.20	0.02-0.022
"	-	-	100	0.026
Felt (dark grey wool)	0.15	-	40-70	0.056-0.063

Heat Conductivity of Materials (continued)

Materials	Volumetric weight	Moisture % of volume	Mean temperature in C.°	Coeff. of heat cond. cal/m/h/°C
Felt (common)	0.30	-	0	0.04
Felt soaked in asphalt	0.88	-	30	0.086
Felt treated with asphalt	0.88	-	0	1.05
Granite	-	-	-	3.00
"	2.51-3.05	-	20-40	2.7-3.5
Hard graphite	1.58	-	50	38.0 (?)
" "	1.9-2.3	-	20-40	4.2
Gravel, washed, pebbles 3-8 cm	1.85	0	0-20	0.29-0.32
Fine clayey sand	2.02	14 by weight	20-40	2.0
Fine river sand	1.52	0	0.20	0.26-0.28
" " "	1.64	6.3% by w.	20-50	0.97-0.99
Fine quartz sand	-	3% " "	20-40	0.05
Soil (sand, sandy clay, gravel) in the open air	1.90	-	0	2.0
Same, beneath buildings	1.80	-	0	1.0
Fill of dry sand	1.6	-	0	0.75
Dry dirt	-	-	-	0.12
Moist dirt	-	-	-	0.58
Dirt with coarse gravel	2.04	-	0-70	0.43-0.50
Sandstone, not case-hardened	2.26	Normal	10-40	1.33-1.53
Same after drying for 6 months	2.25	-	10-30	1.08-1.14
Limestone	2.56	-	100-300	1.08-1.14
"	2.00	-	0	1.00
" fine-grained	1.66	-	20-40	0.58
" coarse-grained	1.99	-	20-40	0.80
Quartz, parallel to the long axis	-	-	0	11.52
" , at right angles to the long axis	-	-	0	6.12
" , fused	-	-	100	0.12
Oak boards, at right angle to layers	0.83	Dry	0-15	0.17-0.18
Oak boards, parallel with layers	0.82	"	12-50	0.30-0.37
Oak, air-dried	0.7-1.0	-	20-40	0.18-0.31
Oak, green	0.9-1.3	-	"	"
Pine boards, at right angle to layers	0.55	Dry	0-50	0.12-0.14
" " , parallel to layers	0.55	"	20-25	0.30-0.32
Pine, air-dried	0.3-0.8	-	20-40	0.14-0.31
Pine, green	0.4-1.1	-	"	0.14-0.31
Sawdust	0.22-0.25	-	0-20	0.06-0.08
Wood ashes	-	-	-	0.06
Slag, from boilers, coal	0.7-0.8	0	0-20	0.12-0.14
Slag from open hearth furnaces	0.785	-	0	0.14
Peat	0.13	-	0-50	0.03-0.04

Other materials with a very low coefficient of heat conduction are: Diatomite, pumice, pumiceous tuff, and scoriaceous and vesicular lava.

DESTRUCTIVE ACTION OF FROZEN GROUND

Swelling of Ground

By swelling of ground is meant the volumetric and other changes of surficial deposits due to alternating seasonal freezing and thawing. Strictly speaking, swelling of ground consists of two phases which from the standpoint of their effect on the earth's surface and buildings consist of swelling and settling or subsidence. It is also necessary to make a distinction between the term swelling, as used by the Russians, and the term frost-heaving which is generally used in America. Frost-heaving denotes a force upward and is only a part of the surficial manifestation of the swelling, whereas swelling refers primarily to the increase in volume with the directional components upward, sideways, and according to some authorities, even downward.

Swelling of ground may affect large areas which may be measured in kilometers or it may be limited to a few square meters. A uniform and widespread swelling usually causes little damage to roadbeds and other structures, but the irregular or differential swelling is very destructive. The three major factors which cause differential swelling are:

1. Unequal distribution of load (in a structure).
2. Differences in the texture of ground.
3. The differences in the amount of water available for the growth of ice.

Swelling of ground can take place under a variety of conditions, but is most common:

1. Where the upper layers are relatively impermeable but saturated.
2. Where water-bearings strata or fissures are present within the active layer.
3. Where the upper layers, such as gravel or sand, are permeable but are underlain by an impervious layer such as permafrost, compacted clay or bedrock.

Field observations and laboratory tests show that the swelling of ground is caused by:

1. Hydrostatic pressure of water.
2. Increase in volume when water is converted into ice.
3. Force of crystallization of ice.
4. Any combination of the above three factors.

According to Sumgin, pressure in freezing ground is caused by either increase in volume when water is converted into ice or by the force of crystallization of ice. The initial pressure may cause the masses of as yet unfrozen water and ground to move laterally in one or the other direction as well as upward. Thus the ground that offers least resistance to this pressure will be forced to move in the direction of lesser resistance, producing either a vertical heave or lateral dislocation.

Taber claims that the pressure effects which accompany the freezing of ground are due entirely to the force of crystallization of ice. According to Taber, "heaving is upward because that is the direction of heat conduction rather than because it is the direction of least resistance" and "the pressure is developed only in the direction of crystal growth, which is determined chiefly by the direction of heat conduction and the availability of water."

In the engineering practice two types of swelling of ground are recognized by Lukashev: superficial swelling and deep-seated swelling.

Superficial swelling usually attains a height of from 5 to 10 centimeters. It is caused by the freezing of meteoric water in the roadbed of a railroad. This water usually accumulates in the, so-called, "shoulder troughs" which are the compacted depression produced by the weight of heavy traffic. Superficial swelling may also be caused by the silting of road metal or by the settling of ballast. In railroad cuts, due to a greater density of the surrounding ground, the accumulation of water immediately below the roadbed is usually greater than in fills and the swelling may reach a height of 10 centimeters, whereas in fills it seldom exceeds 3 to 5 centimeters. Moreover, in fills, the swelling of ground may be partly directed sideways.

Deep-seated swellings are more pronounced in cuts indicating that they are caused by the freezing of ground-water that percolates beneath the roadbed. They average in height from 30 to 40 centimeters but frequently exceed 50 centimeters. Deep-seated swelling is strongest where permafrost table is close to the surface and merges with the frozen active layer. This condition commonly creates high hydrostatic pressure which is responsible for the intense swelling.

In general, it has been observed that the thicker the active layer the more pronounced is the heaving of the structures. It has also been established that the swelling begins earlier in cuts than in fills. Fills, in turn, show swelling before ungraded places are affected. This is explained, at least in part, by the fact that cuts receive less solar radiation than fills and that fills have greater surface exposed to cooling than ungraded parts of terrain. In the spring the subsidence of the swelled ground proceeds in the reverse order.

The most conspicuous manifestation of the swelling of ground is represented by frost-mounds. (Fig.27). Mounds produced by frost action vary widely in size, structure, origin, and duration. The available information is as yet inadequate to arrange them systematically according to their morphology and origin.

Scores of different terms have been proposed by different authors to designate the various types of mounds. Indiscriminate use of these terms by subsequent workers has caused much confusion and has made it very difficult to analyze and comprehend the true nature of these phenomena from their, often incomplete, descriptions. There are undoubtedly many different types of mounds in existence whose origin, morphology and history are quite distinct but in only a very few cases have these been thoroughly studies and objectively described. To minimize further confusion it is proposed that the term frost-mound be applied to all mounds produced by frost action, unless their specific character, origin and structure are known. Frost-mound will thus serve as a family name and within it may be provisionally distinguished the following special types:

Pingo (Fig. 28) - a large mound which may attain a height of about 100 meters and a perimeter of over 1,000 meters. Pingos are of many years duration but the factors which control their life-cycle have not yet been determined. According to Porsild, a pingo is caused by a progressive downward freezing of a body or lens of water or of semi-fluid mud. Pingos are known to occur along the Arctic coast in flat, poorly drained areas near a river delta or in an old lake basin. Pingo-like structures, locally known as "boolgunyakhs" or "hydrolaccoliths" have been also described from the vicinity of Yakutsk and in Transbaikalia, Siberia. The summit of a pingo is usually broken by radiating fissures from which may issue potable water.

Frost-blister is a mound or an upwarp of ground produced by a localized hydrostatic pressure of ground water. (Fig. 29).

According to accounts of observers[2/], the presence of water in a growing frost-blister can be felt through the springy feel beneath one's feet. Hydrostatic pressure may exceed the elastic limits of the overlying crust of frozen ground and the frost-blister may rupture with a loud sound like a gun-shot. Some water usually gushes out of the broken blister and freezes before it has a chance to spread very far. The remaining water freezes within the blister forming a kernel of ice in the middle of the mound. The rupture of a frost-blister can be anticipated and the pressure within it can be relieved by puncturing it with a crow-bar.

The ice within a frost-blister may not melt during the following summer and as a result the frost-blister mound may last longer than a year.

Frost-blisters usually form along sloping ground and may shift their position from year to year. They are usually of small or moderate size, seldom exceeding 8 meters in height.

2/ Personal communication from C. C. Nikiforoff.

End view of seasonal frost-mound. The willows protruding through the sides of the mound, with part of the surface soil in which they grew, were dislodged by the upheaval.

(From Porsild)

FIG. 27

Frost-mound (PINGO) near Tuktuayaktuk on the Arctic Sea coast of the Mackenzie delta showing the radiating rupture of the summit. This pingo is 40 meters high.

(From Porsild)

FIG. 28

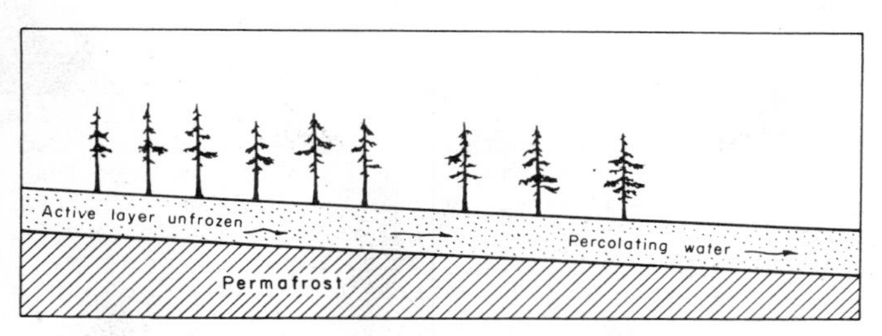

Ground water in active layer percolates freely down slope.

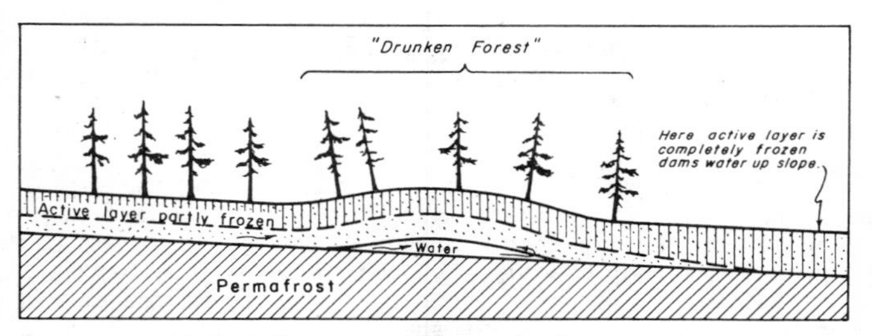

Hydrostatic pressure pushes up surface ground.

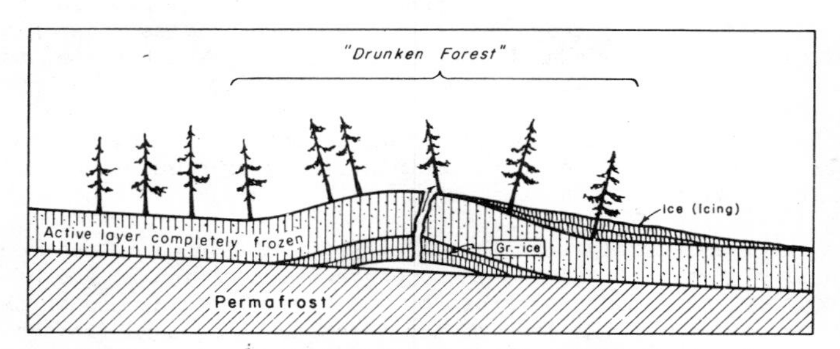

Mound is ruptured by hydrostatic pressure (and crystallization of ice ?). Water freezes forming icing and ground-ice. Occasionally a hollow space is left in the core of a mound.

DIAGRAMS ILLUSTRATING FORMATION OF FROST-BLISTER

Modified after Nikiforoff

FIG. 29

Similar mounds have been described by various authors as soil-blister, ice-mound, earth-mound, gravel-mound, etc.

Icing-mounds (Figs. 30 and 31) are, for the most part, composed entirely of ice. They are formed by the freezing of successive layers of water which may issue either from the ground or from a fissure in river ice. Locally the growth of an icing-mound is preceded by a bulge in river ice or in a surficial layer of soil. They are thus similar in origin to frost-blisters, the main distinction being that the source of their water is not as limited and for that reason they issue water for a considerably longer time than do the frost-blisters. However, admittedly, no sharp distinction can in some cases be drawn between these two types of mounds.

Icing-mounds measure from about a meter to more than ten meters in height and width and are commonly surrounded by a more or less extensive sheet of layered ice - the icing ("naled'" or "aufeis"). It does not necessarily follow, however, that all icings have an icing-mound near the place of issue of water. Some icings may spread over a considerable area without having a prominent mound at the source of water.

Most icing-mounds are annual, forming during the winter, but in exceptional cases large icing-mounds do not melt completely during the warm season and so may last a number of years. Icing-mounds may appear from year to year at exactly the same place or they may spring up in slightly different places. Occasionally icing-mounds make a sudden appearance at a locality where they have not previously existed.

Peat-mounds are small hummocks which are commonly present in swamp or tundra country. They form surface irregularities which give rise to the names "hummocky swamp", "hummocky tundra", and "spotted tundra". Peat mounds usually last longer than one season.

Little is known about their origin and structure. Peat-mounds in the Kola Peninsual were found to contain lenses of ground-ice within the moisture saturated peat and moss.

It seems likely that the origin of peat-mounds is due to a localized swelling of ground caused by the segregation of ice in ground which is unequally insulated from frost by the variable thickness of the vegetative cover. In the less protected spots the ground freezes sooner and at a more rapid rate than in the adjacent and better insulated areas. The ice that forms in the places that freeze first will grow by drawing moisture from the adjacent ground so that when the better insulated ground freezes, the amount of available moisture in the ground will be considerably less. As a result the bare or less protected places will remain in a slight relief as compared with the adjacent areas which are protected by a thicker cover of peat or moss.

Side view of icing-mound in Siberia.
The mound is about 50 meters long, 16 meters wide and 7 meters high.

(From L'vov)

FIG. 30

End view of the same icing-mound.
A group of tilted trees is referred to as "drunken forest". (From L'vov)

FIG. 31

The redistribution of ground moisture during freezing may also, at least in part, be due to the difference in texture of the ground while the vegetative cover may remain fairly uniform over the entire area.

Peat-mounds have been also described as "palsen" (singular - pals).

Relation of Texture to Swelling. The relative intensity of swelling and heaving of ground during freezing is determined by the texture of the material. Some grounds swell very strongly whereas others, even under the most adverse conditions practically do not swell at all. In this regard three main groups of grounds can be distinguished:

1st group: Grounds that do not swell:

a. Solid bedrock.
b. Ground composed of coarse fragments (pebbles and boulders) with the interstices filled with smaller fragments of the same material.

2nd group: Grounds that swell little:

a. Gravel and coarse sand with the admixture of fine particles (clay or silt), moist or saturated with water but without available supply of additional water.

3rd group: Grounds that swell the most:

a. Gravel and sand mixed with clay and silt, moist or saturated with water and with access to additional supply of water.
b. Sandy clay, clayey sand, and clay.
c. Fine sand, sandy silt, silt, and slud (wet mud).
d. Peat.

Grounds "b" and "c" of the third group are most susceptible to swelling, particularly when saturated with water.

In fine-grained deposits capillary action plays an important role in the swelling and heaving of the ground. Water freezes at the usual temperatures in large capillary voids but in very fine capillaries water does not freeze at temperatures as low as -17° C. and even down to -78° C. Water which freezes in the larger capillaries draws the unfrozen water from finer capillaries and this water is continually being added to the growing crystals of ice. In this manner the volume of ice continues to increase as long as there is a supply of water. This process explains the strong swelling which usually takes place in silty and clayey grounds underlain by water-bearing strata.

Unindurated grounds with only a small amount of capillary voids and with no access to a supply of free interstitial water from below or from the sides do not undergo any significant swelling. Thus in sandy ground the swelling is very slight and is almost entirely due to the crystallization of water in the interstices at normal freezing temperatures. In sand the amount of unfrozen water which is drawn from the capillaries by the crystals of ice is so small that it does not contribute to the general increase of the volume of ground when frozen.

The conclusion that silty and clayey grounds with high capillarity suffer far stronger swelling than the sandy grounds is well borne out by field observations and by experiments. (Fig. 32).

Taber's experiments show that "in indurated clay, or clay that has been thoroughly consolidated artificially, the layers of segregated ice are clear, for the most part, and very sharply separated from the frozen clay. The total thickness of these layers, as close as can be measured, is the same as the amount of surface uplift". This offers further support to the conclusion that, at least in some cases, the swelling of clayey ground is caused primarily by the growth of ice layers which draw on the supply of additional water from the capillaries of the adjacent water-saturated ground. This condition under which an additional supply of water is available during the process of freezing is known as the <u>open system</u>. It is contrasted with the <u>closed system</u> under which the freezing ground has no access to an additional supply of water from without.

The swelling of different grounds is graphically shown in Fig. 33.

<u>Relation of Hydrologic Conditions to Swelling</u>. Field observations and experiments show that the character and magnitude of the swelling of ground depend on the following hydrologic conditions:

1. Amount of water in the ground.
2. Whether or not additional supply of water is available - whether the condition present is an open or a closed system.
3. The rate of flow of additional water to the place of swelling.
4. Whether or not conditions are present to create hydrostatic pressure.

Normally, grounds swell only if the water they contain is in excess of the <u>critical moisture content</u>[3]. If, on the other hand, their moisture is below this critical point and if no supply of additional water is available (closed system), little or no swelling will take place. Grounds which contain water in an amount exceeding the critical moisture point will swell, regardless of whether or not an additional supply of water is available. Additional water will increase the intensity of swelling and heaving.

3/ Critical moisture content is defined as the maximum amount of interstitial water which, when converted into ice, will fill all the available pore space of the ground.

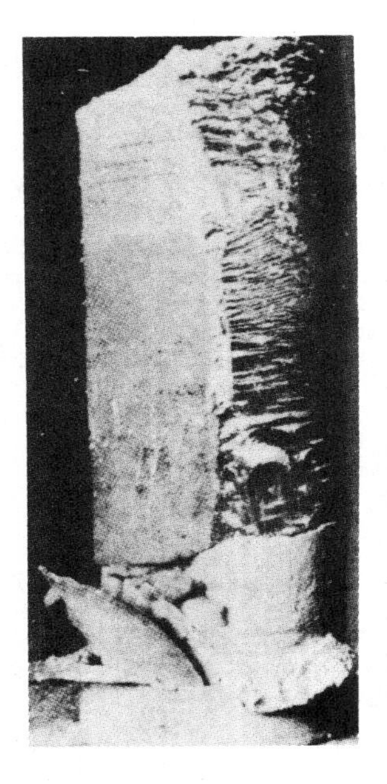

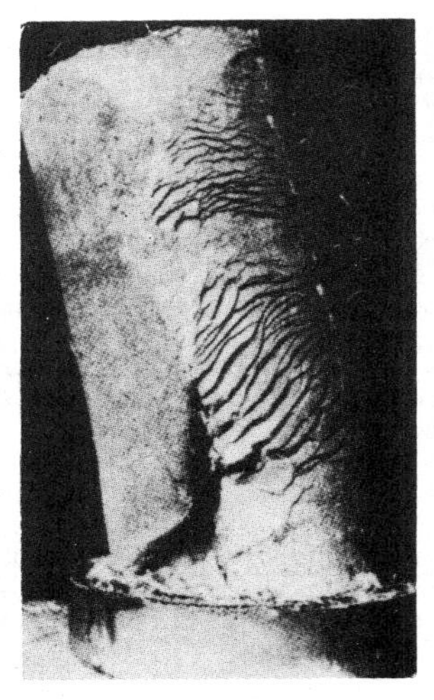

A. B.

Experimental freezing of sand (left) and clay (right) in an open system.

A. Frozen cylinder, half sand and half clay. Much segregated ice in clay but not in sand.

B. Differential displacement of cylinder due to segregation of ice (ice lenses) in clay but not in sand. Cavity caused by dislodgment of dry sand.

(From Taber)

FIG. 32

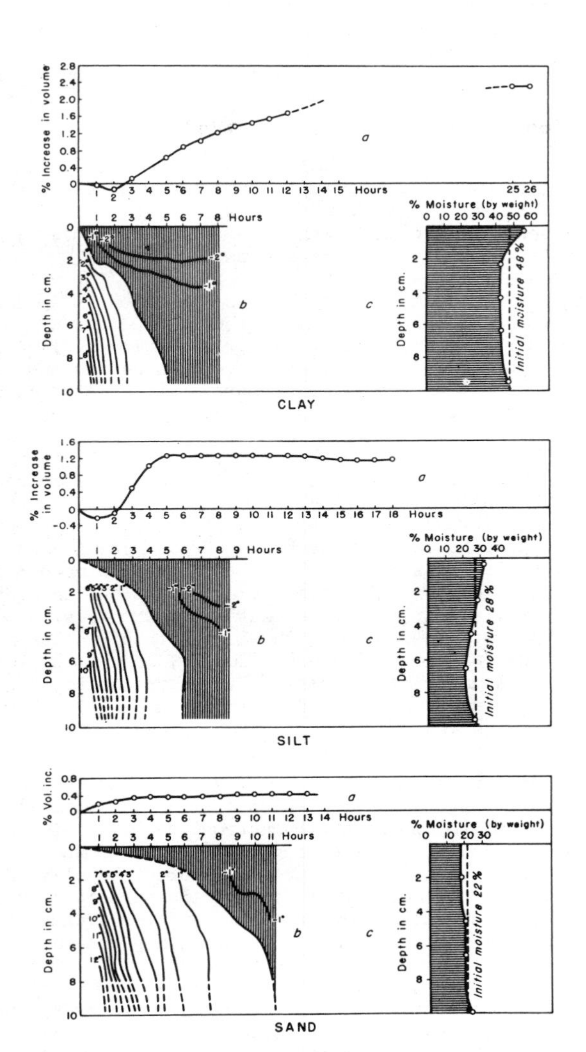

Swelling of grounds on freezing from the top.

a. The curve of swelling ground.
b. Temperatures.
c. Distribution of moisture in the frozen sample.

(After Vologdina)

FIG. 33

Theoretically no swelling should take place if the amount of water in the ground does not exceed the critical moisture content, but in practice a slight swelling is usually observed. This is explained by the fact that the distribution of water in the ground is probably such that in some places ice fills only a small part of the available pore-space, while elsewhere it is in excess.

Experiments by Andrianov also show that the swelling of ground in a closed system, is less than the amount to be expected had all the water contained in the ground turned into ice. It is believed that this difference arises from the fact that some of the water represents a fixed moisture which normally does not freeze.

Taber's experiments with an open system show "that the frozen clay between the ice layers contains the same percentage of water as clay below the depth of freezing". In Taber's opinion "this observation confirms the conclusions that heaving .... is due to the formation of ice layers and that the freezing of interstitial water causes practically no uplift." This opinion, however, is not shared by most Russian authors.

A marked swelling of ground usually takes place where a considerable amount of water is added to the ground during the process of freezing. When downward freezing reaches the capillary fringe, water is drawn through the capillaries by the frozen ground and is added to the already formed mass of ice. The height of the mound thus formed on the surface of the ground is directly dependent on the supply of available water. Taber maintains that the ultimate size of the mound and "the pressure developed through the growth of ice crystals" are "limited by the failure of the water supply, and not by inability of crystals to grow under great pressure". Taber admits, however, that the increased pressure upon the freezing ground noticeably retards the segregation of ice into thick and sharply separated layers.

A very destructive type of swelling of ground is due to hydrostatic pressure of ground-water which is trapped during the winter freezing between the impervious permafrost below and the frozen active layer above. This type of swelling is most common where the permafrost table is close to the surface and where the active layer freezes clear through and ultimately merges with the permafrost. Under this condition bodies of shallow ground-water with no avenue of escape are squeezed, as in a vise, and cause the overlying ground to bulge and crack open.

The mechanics of this type of swelling is complex, for in addition to the hydrostatic pressure, the forces of crystallization of ice and of pressure due to the expansion of water when passing from a liquid state into ice all take part in variable proportions in producing the total effe

The nearer the ground-water is to the surface the stronger is the swelling of ground. Voislav has also observed that the crest of the swelling is commonly situated directly over the main seepage of ground-wat

In summary:

1. Swelling of ground when additional water is available (in an open system) is due to:
   a. Force of crystallization of ice.
   b. Capillary lift of water.
   c. Hydrostatic pressure of water.
   d. Increase in the volume of water during freezing.
2. The amount of ice formed during the swelling of ground is determined by:
   a. Size and shape of ground particles.
   b. Size and amount of voids.
   c. Amount of water available.
   d. Rate of freezing.
   e. Surface load or resistance to heaving.
3. Favorable conditions for swelling of ground are:
   a. Occurrence of ground-water at shallow depth in areas free of permafrost.
   b. Occurrence of permafrost at shallow depth.
   c. Fine-grained texture of the ground.

Relation of Freezing to Swelling. The main factors are:
1. The thickness of the active layer, which is commonly the same as the depth to the permafrost table, and
2. The depth to the permafrost table.

If the active layer is relatively thin, as for example, along the Arctic Coast, there is little likelihood of any considerable swelling of ground. Farther south in the permafrost region the summer thaw extends down to a depth of from two to four meters and the surface of ground is usually quite hummocky. As has been already pointed out the relative sizes of these hummocks or mounds depend on the texture of the ground and the hydrologic condition of the area.

Where permafrost begins at a considerable depth below the surface and where the freezing of active layer does not reach that level the intermediate ground remains unfrozen forming a talik. These taliks tend to minimize the swelling of ground and where they attain a thickness of from 4 to 5 meters the surface effect of swelling is practically nil.

Swelling of ground takes place primarily under the following conditions:

1. When the ground is composed of silty, fine sandy or clayey deposits, or peat.
2. When these deposits are moistened by capillary or ground-water, and
3. When the freezing of the active layer causes the freezing of the shallow ground-water which is backed against the underlying permafrost or impervious unfrozen ground.

Settling and Caving

When ground of fine texture thaws it becomes more or less plastic and frequently forms a mud-like material of paste consistency which is called slud. (Figs. 34 and 35).

It is thought desirable to keep a separate designation for this mud-like material instead of employing the term mudflow, as had been suggested by some writers. The mudflow is a very definite geologic phenomenon arising from excessive wetting of the surficial debris or due to percipitation or to rapid melting of snow, as for example, on the slope of volcanos (volcanic mudflow).

If thawed ground is plastic enough, it flows or oozes out from beneath a load (building or other structure) causing damage to the structure by settling. Most commonly a marked settling of ground is due to the melting of layers or wedges of ground-ice, especially where these are close to the surface of the ground.

As with swelling, the magnitude of settling or caving depends on:

1. Composition of the ground.
2. Amount of ice.
3. Depth of seasonal thaw.
4. Other factors which affect the normal depth of seasonal thaw.

The rate of settling depends primarily on the rate of thawing, which, in turn, depends on the temperature gradient of the ground, specific heat of the permafrost and of active layer, type of construction, and dimensions and orientation of the structure.

Other conditions being equal, even a very light structure tends to settle more on the south-facing side than on its northern side. The ground along the south wall receives the normal amount of solar radiation plus the reflection from the wall of the building and will thaw to a greater than normal depth, whereas the ground on the shaded north side, will be affected in just the opposite way. A simple board fence, running east and west produces a noticeable difference in the temperature of the underlying ground on the two sides because the amount of solar radiation received by the ground is much greater on the south side of the fence. (Fig. 36).

Settling is frequently caused by excessive thawing of frozen ground due to faulty planning and construction of a building. A miscalculation of the amount of heat that is likely to be generated by a building and transmitted into the underlying frozen ground may result in a deeper thaw of ground than was expected and may cause irrepairable damage to the building. (Fig. 37).

Thawed ground of mud consistency (slud), Fairbanks district, Alaska.

(From the U.S.Geol.Survey files)
(Photo by Taber)

FIG. 34

Thawed ground of mud consistency (slud),McKinley Park,Alaska.

(From the U.S.Geol.Survey files)
(Photo by Taber)

FIG. 35

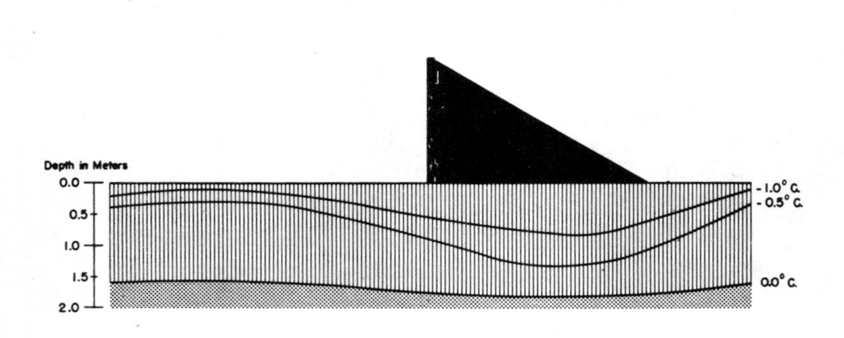

Annual Ground Isotherms Affected by the Shadow from an East - West Fence.

FIG. 36

(From Tsytovich and Sumgin)

Landslide, Slump, Creep, and Solifluction

Landslide, slump, creep and solifluction are phenomena which consist of downward falling, sliding or flowing of surface ground material. Following Sharpe, landslide is defined as "the perceptible downward sliding or falling of a relatively dry mass of earth, rock, or mixture of the two". Landslides usually occur along steep slopes and the rate of their movement is variable. A slump is similar to a landslide but is "usually with backward rotation on a more or less horizontal axis parallel to the cliff or slope from which it descends". The term creep refers to a "slow downslope movement of superficial soil or rock debris, usually imperceptible except to observations of long duration". Solifluction refers to a slow downslope movement of water-saturated masses of surface ground (blocks, slabs, and gravel with finer particles filling the interspaces). To this may be also added a mud-flow which usually has a higher content of water and moves more rapidly.

The above processes are very common in the permafrost region and their destructive effect on buildings, roads, etc. is familiar to all.

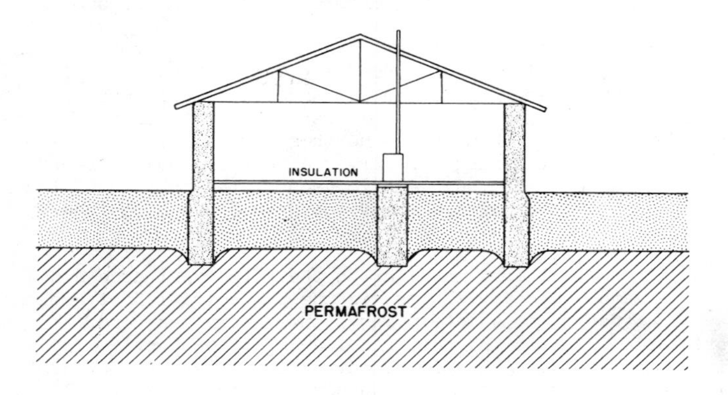

Maximum permissible lowering of the permafrost table beneath a foundation.

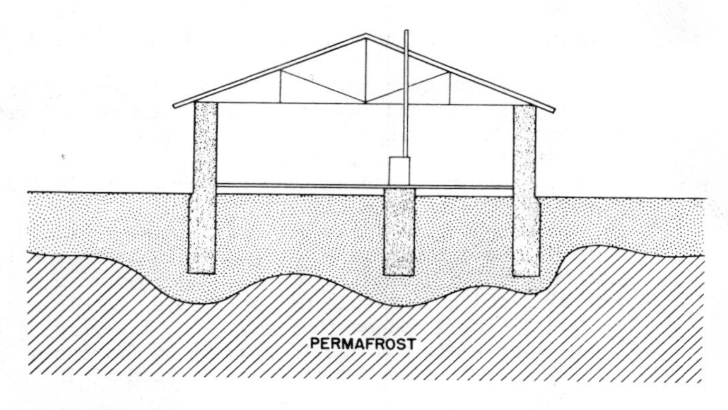

Insufficient insulation or excessive heat transfer into ground caused thawing of ground. If thawed ground has low bearing strength the building is likely to settle and be damaged.

THAWING OF PERMAFROST BENEATH A BUILDING

FIG. 37

In the permafrost region, however, ground which is susceptible to sliding or flowing may be frozen solid during the winter and may remain stable until it is thawed and rendered plastic in the summer.

It is therefore easy to see that in the permafrost region the thermal regime of layers of ground close to the surface plays an important role in the areal distribution and relative intensity of these processes. Landslides and slumps are more common on the south-facing slopes of mountains than on the north sides, a fact which should be taken into account during the survey of a road route. Destructive slides and slumps have been also caused by the melting of the permafrost where the thermal regime of the ground was disturbed by human activity.

Another condition, peculiar to the permafrost area and a common cause of landslides and slumps, is the concentration of ground moisture at shallow depth, immediately above the impervious permafrost or over the frost table prior to the complete thawing of the active layer. The moisture content of this ground may be in excess of 100% of its dry weight. Such ground is highly susceptible to flowage. Sliding of unfrozen ground along the permafrost surface is the commonest form of landslide in the permafrost area. Much difficulty is also experienced with the flowage of ground in roadcuts and other excavations.

Safeguarding measures against landslides commonly include:

1. Control of the hydrologic regime of the area (prompt and thorough repair and upkeep of water lines), diversion of seepages, and drainage in general.
2. Prevention of deforestation.
3. Planting of trees and grasses on the slopes that may possibly develop landslides.
4. In choosing a line for a railroad, places should be avoided where there is a sharp drop in the permafrost table near the shoulders of the projected road bed (Fig. 38).
5. Quarries and pits should not be located in close proximity to the railroad bed as they may cause lowering of the permafrost table and create conditions favorable for a slide of the road bed in the direction of the excavation.
6. Artificial freezing of ground has been successfully employed where steep banks were endangering the excavation operations.

Road built over a shoulder of permafrost table is susceptible to landslides and deformation by differential heaving.

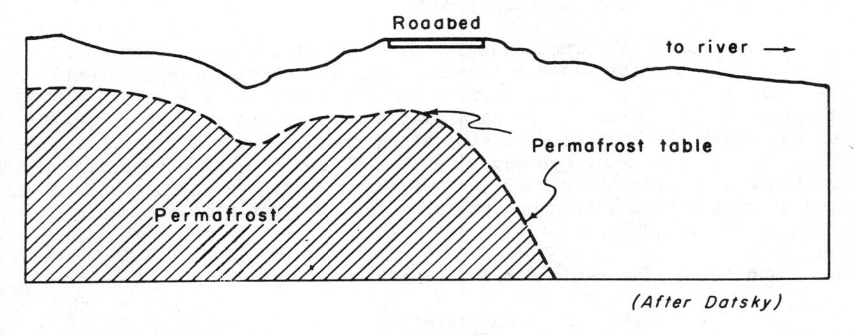

(After Datsky)

Profile of the permafrost table near a river

FIG. 38

Icing

The term icing ("glaciers", aufeis, in Russian "naled'") is proposed for a mass of surface ice formed during the winter by successive freezing of sheets of water that may seep from the ground, from a river, or from a spring. Icing is thus a coating or a crust of ice which has, at least a superficial, analogy with the icing on a cake.

The term "glaciers", applied by the natives of Alaska to these sheets of ice, is most inappropriate and is likely to cause confusion. The surficial sheets of ice discussed here have no connection with the true glaciers.

Icings commonly form irregular sheets or fields, mounds (icing-mounds) or irregular incrustations along slopes (excavations, river banks, etc.). The surface of the icing is usually very uneven and when it spreads over a road it may make the road impassable. Ground-water seeping out of the fractures of a rock may result in a formidable ice incrustation of damaging proportions so that serious difficulties may be encountered in road cuts made in fractured bedrock.

Icing may form prominent mounds - icing mounds (Figs. 30, 31, and 39) - which may attain the height of 20 meters and may have a horizontal dimension of 50 meters or more.

River icing is formed when the freezing of a shallow portion of a stream impedes the flow up-stream. This increases the hydrostatic pressure of water above the barrier and the water is forced to break through to the surface. The water may emerge from a fissure in the ice or may reach the surface as a seep along the bank of the river. Occasionally the break-through occurs at a weak spot in the floodplain some distance away from the stream. River icing therefore may not only cover the channel of the stream but may also "spill" over a considerable portion of the floodplain.

Conditions favorable for the formation of icings are:

1. Presence of ground-water in the active layer.
2. Low temperatures of the air and only thin cover of snow during the early part of the winter (December, January).
3. Proximity of the permafrost table to the surface of the ground.
4. Thick cover of snow during the latter part of the winter.

Icings are formed when the winter freezing of ground penetrates down to and merges with the permafrost. The ground-water in the active layer is forced to the surface along a path of least resistance, spilling over the surface and freezing.

Icings are more likely to develop along a mountain slope where the strata dip in the same direction as the slope of the mountain. They are less frequent along a slope which is underlain by strata that dip into the mountain.

Icing-mound near the station Omootnaya, Siberia.

(From L'vov)

FIG. 39

In southern Siberia icings are most intense and attain their maximum spread in March when the snow-cover is also at its maximum. Snow acts as an insulating blanket preserving in the underlying ground the low temperatures which were caused by the December and January frosts.

The insulating effect of snow is strikingly illustrated by the following observations of P. I. Koloskov in 1915 at Gosh, Siberia:

	Jan.	Feb.	March
Thickness of snow	19 cm.	24 cm.	38 cm.
Temperature on the surface of snow	-47.8°	-40.5°	-31.0° C.
Temperature beneath the snow	-19.2°	-14.9°	-8.9° C.

A slight difference in the thickness of snow may have a far-reaching effect on the behavior of an icing. V. G. Petrov states that one large icing which has been appearing regularly for many years failed to form altogether during a winter with an unusually heavy snowfall. How a slight difference in the thickness of snow can affect the temperature of the ground is shown by M. I. Sumgin at Bomnak Permafrost Station in Siberia:

	Dec. 1913	Dec. 1914	Dec. 1918
Thickness of snow	5 cm.	24 cm.	30 cm.
Temperature of the air	-5.2°	-4.7°	-4.7° C.
Temperature of the ground 1.5 meters below the surface	-2.33°	-0.29°	+0.20° C.

In areas with heavy snowfall during the early part of the winter rive and ground icings usually appear late in the winter or early spring, or may not appear at all. On the other hand in areas with less than half a meter or no snow at all, icings are likely to appear as early as Decembe

During the summer some large icings do not melt completely leaving remnant ice-fields which the natives in Siberia call taryns.

River icings are especially destructive to bridges and roads. Seepages of water may unexpectedly develop beneath or near the buildings where the ground fails to freeze over during the winter. The effect of icing formed under these conditions is vividly illustrated in Figs. 40 - 45. To prevent the destructive action of icing it is necessary to divert the flowing water or to intercept and drain the ground-water which feeds the icing.

Icings caused by heated buildings.

Thawed ground under an inhabited house caused a sudden outpour of water which quickly froze, filling the house with ice.

(From L'vov)

FIG. 40

Ground-water broke through the thawed ground beneath the bath-house, quickly filled the house with ice.

FIG. 41 (From L'vov)

Icing formed within a bath-house.

Water seeped in from beneath the floor of bath-house, filled the entire building and froze. The house was torn down exposing the ice "mold".

(From Tsytovich and Sumgin)

FIG. 42

House almost completely engulfed by icing.

FIG. 43 (From Tyrell)

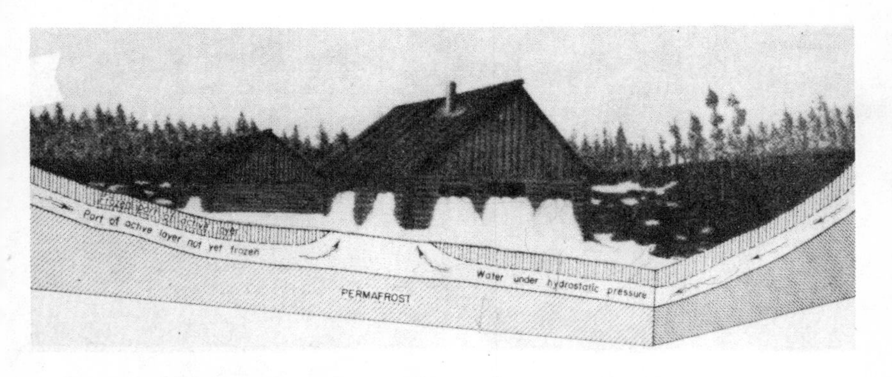

Water seeping through unfrozen ground beneath an inhabited house fills the building with ice.

FIG. 44

An overturned barrel left standing in a yard started an icing which filled the lower half of the barrel and spilled through the plug-hule in the middle. The icing spread over the greater part of the yard and flooded some buildings.

FIG. 45

Thermokarst

The term karst, as typified at Karst on the Adriatic Sea, refers to an uneven topography with short ravines, sink-holes, funnels, and caverns which are produced in a limestone terrain by the solvent action of water. The term thermokarst is used by the Russian scientists for a similarly irregular land surface in the permafrost region where the depressions and caverns are the result of thermal action - the melting of ground-ice.

Thermokarst land forms are commonly produced by forest fires, grass fires, deforestation, and stripping of the surface by man. Removal of the vegetative cover results in a more intense solar heating of the ground, melting the ground-ice and causing settling or caving of the ground.

In a thickly forested region partial deforestation will cause a more rapid development of thermokarst phenomena than in a sparsely forested area.

Thermokarst processes are more intense where the natural balance of the frozen ground is disturbed through the action of man, as, for example, in farming or in road building.

Development of thermokarst features near a populated area is commonly caused by the stripping of the protective layer of peat or sod for building purposes, by poorly constructed drainage ditches, or by trampling of the ground by people or cattle. A badly planned disposal of sewage, especially containing hot water and kitchen refuse, may initiate a thermokarst process which may rapidly spread beyond control and seriously endanger the building.

The relative intensity of thermokarst processes depends on several factors of which the most important are:

1. Composition of the ground. Ground that contains a considerable amount or entirely consists of carbonate, gypsum, anhydrite, rock-salt, and ice is especially susceptible to the thermokarst processes.
2. Structural relations of ground that is susceptible to the thermokarst processes.
3. Climatic factors.
4. Hydrologic and hydrogeologic factors (circulation and regime of surface and ground-waters and their mutual relations).
5. Condition of the soil.
6. Geomorphic factors.
7. Human activity.

In the study of thermokarst phenomena it is important to determine what role each of these factors plays individually and in combination with other factors. The investigator should also evaluate the already developed phenomena in order to predict the extent and direction of their future development.

Some of the commonest land forms which are produced by the thermokarst processes are:

1. Surface cracks. Surface cracks produced by the thermokarst processes differ from the frost cracks in that they are usually wider, deeper, and longer. They usually are formed in the summer and, as a rule, last through the winter, whereas the frost cracks form in the winter and usually seal up in the summer.

   Surface cracks frequently precede landslides, slumps, and tears and provide a passage-way for the outflow of wet mud-like ground called slud.
2. Cave-ins and funnels. Cave-ins and funnels are depressions which may be either dry or filled with water. They normally do not exceed 10 meters across. Their shape may be regularly conical, cylindrical or quite irregular and their sides may be smooth or uneven and traversed by numerous cracks.
3. Sinks and saucers and shallow depressions varying from a few meters to several hundred meters in width and from a decimeter to several meters in depth. Their slopes converge in the center and are usually very gentle or slightly concave. In larger depressions the bottom may be somewhat uneven with smaller depressions or mounds. Occasionally the slopes are benched or form a steep embankment. The northern slopes are usually gentler than the southern.
4. "Valleys", gulleys, ravines and sag basins are as a rule of large dimensions and attain a size of several kilometers. Their surfaces are frequently modified by the smaller features described above.
5. Cave-in lakes, windows, and sag-ponds represent later stages of the depressions listed above. When these become either temporarily, periodically or permanently filled with water, the thermokarst origin of such lakes is usually recognized by the following evidences:

   a. Submerged trees, shrubs, and grasses are the same as those which grow on the adjacent dry ground.
   b. Submerged sod.
   c. Cracks in the bottom of the lake.
   d. Little if any silting of the basin.
   e. Absence of well established aqueous plants (hydrophytes).
   f. Presence of open cracks along the shore.

   In time a cave-in lake becomes completely matted over with moss and peat showing occasional buried tree trunks and stumps.

ENGINEERING PROBLEMS PECULIAR TO AREAS OF PERMAFROST

Introduction

The Arctic and Subarctic Regions are notable for their vastness, scarce population, and general hardships arising from lack of roads and extreme cold. Endless miles of swampy tundra or muskeg, broken by isolated patches of dry ground with scraggly spruce or hardy birch, aspen, and alder, form a typical landscape of the north. Except for the extreme north, mountain slopes are clothed with denser and healthier looking forests but even these gradually diminish in vigour, as one ascends to higher elevations, and eventually give place to moss-covered alpine tundra or bare, rock-strewn slopes and summits. River floodplains and flat coastal areas abound with lakes and aimlessly winding rivers divided into numerous channels, some of which with time become partly silted and separated from the main stream, producing elongate crescent-shape lakes known as "ox-bows". Over most of this area the ground, a short distance below the surface, is permanently frozen locally down to a depth of several hundred feet and does not thaw out even during the warmest summers.

Civilization has been slow in penetrating this northern wilderness heretofore dared only by an occasional lone trapper or prospector. But with the recent progress in aviation and due to strategic contingencies of the present war many remote corners of the North have become sites of active construction and settlement by military and civilian personnel.

Almost unsurmountable difficulties were experienced by the engineers who were building roads, runways, and buildings in this new, severe, and seemingly unfriendly environment. At first an attempt was made to "fight" the natural forces of frost by using stronger materials, more rigid designs, or to resort to periodic and costly repairs, which rarely, if ever, succeed in permanent righting of the situation. It soon became apparent that such a procedure was too costly and in many instances entirely futile; that in the Arctic and Subarctic Regions the usual engineering methods and practices cannot be successfully employed and that the engineering principles had to be adapted to, and reckoned with the prevailing frost conditions.

It has been already amply demonstrated by the costly experience of the Russians that satisfactory results can be achieved if the cyclic (annual) behavior of frozen ground and frost forces are carefully analyzed, understood, correctly evaluated, and are allowed for in all engineering designs in such a manner that they appreciably minimize or completely neutralize and eliminate the destructive effect of frost action.

Once the frozen ground problems are understood and correctly evaluated their successful solution is, for the most part, a matter of common sense whereby the frost forces are utilized to play the hand of the engineer and not against it. A building placed on piles properly anchored in the permanently frozen ground will remain stable and undamaged whereas a structure erected on a shallow foundation is likely to be heaved and damaged, possibly beyond repair. A road built over an undisturbed frozen ground with peat and moss cover left intact will remain stable as if built on a solid bedrock. This method of construction is known as the PASSIVE METHOD. With the passive method of construction the permanently frozen ground conditions are left undisturbed or an additional insulation is provided so that the heat generated by the newly erected structure would not cause any thawing of the underlying ground and thus weaken its stability.

Where permafrost is thin and the ground is of the kind that will have a satisfactory bearing strength when thawed, measures are taken to thaw out the ground prior to the construction and the stresses are computed for the strength of the ground in its thawed state. This method of construction is called the ACTIVE METHOD.

The importance of a preliminary investigation of the permafrost conditions which determine whether the passive or the active method of construction is to be used cannot be overemphasized.

In Russia, where, since the beginning of the century, the problem of permafrost has been a formidable obstacle in the conquest of Siberia, the intensive study of the subject has already brought fruitful results and a general realization that it is more economical to expend hundreds or thousands of dollars on a preliminary survey of the ground than to pay millions of dollars for the repairs and upkeep of ill-planned projects that are doomed to ultimate collapse or failure.

The following brief notes embody some of the recommendations by Russian and other specialists from their recent experience in the construction of buildings, runways, roads, water-supply, and other projects.

Construction of Buildings

General considerations. In planning a building in the permafrost area the first question which the engineer has to decide upon is whether he should employ active or passive method of construction.

As has been shown in the preceding pages, the three main factors which bear upon the relative stability of a building erected on a frozen ground are:

1. Texture and structure of the ground.
2. Temperature of the ground and air.
3. Hydrology of the ground.

Tsytovich and Sumgin, chiefly on the basis of thermal considerations, recommended the following rules for the erection of structures:

1. In the region of permafrost where the temperature of the frozen ground is about -5° the condition of permafrost should be retained - the passive method should be employed.
2. Where the temperature of the ground is from -5 to -1.5° either method may be employed. The choice of one or the other method will depend on local conditions such as saturation of the ground with ice, temperature regime of the structure, and the type of the structure planned.
3. In the southern part of the permafrost region, where the temperature of the frozen ground is above -1.5° the permafrost condition at the site should be eliminated - the active method should be employed. However, there may be some exceptions, as for instance in ground containing a considerable proportion of silt and fine sand, elimination of the permanently frozen condition will convert this ground into a plastic and unstable mass (slud) and it will be necessary to take special measures. Elimination of permafrost conditions - active method of construction - is feasible when the thickness of permafrost is not much in excess of one meter, but it is very doubtful if satisfactory results can be achieved when its thickness is 20 meters or more.

A stable foundation can be built on bedrock irrespective of whether. permafrost regime is retained or eliminated. In general, however, the passive method (retention of the regime) is preferred, as the thawing of ice in cracks (joints) in the rock when the active method is used may result in water seepages beneath the building. In using the passive method the building should be properly designed (insulated) to prevent thawing of ground and lowering of the permafrost table below the base of foundation. If thawing penetrates below the base of the foundation (Fig. 8) the ground will tend to accumulate moisture and may result in a seepage and subsequent icing directly beneath the building or dangerously close to it.

This and similar situations may be remedied by insulation or by proper drainage that will prevent percolation of ground-water or flow of surface water towards the building.

Thawing of permafrost is likely to cause little if any damage to a structure if the texture of the thawed ground is the same as that of the active layer and if the thawed ground does not contain more moisture than the active layer. Care should be exercised, however, with permanently frozen bedrock as some rocks crumble to powder when thawed.

Although laying of foundations within a non-swelling active layer is, in general, fairly safe, it should be avoided with structures of permanent character as there is a possibility of an eventual disintegration of ground due to the periodic freezing and thawing. This may be also true of the thawed parts of permafrost.

Foundations with large bases transmit pressure to a greater depth than do the foundations with properly spaced smaller areas at the base. If the spaces between areas which carry load are smaller than the areas with load, the ground should be examined to a depth approximately equal the width of the entire building, or to at least twice the width of the widest base of the foundation.

Buildings resting on sand are deformed exclusively by the force of frost-heaving which acts only on the base of the foundation. When the active layer becomes completely thawed it may not settle to the original pre-freezing (pre-heaving) level. This difference between the original pre-freezing level and the level of the ground after the thawing and settling is called the residual heaving. Residual heaving, as a rule, does not take place in sand so that after the ground is thawed the buildin settles bodily with the ground to its original (pre-heaving) level.

Construction of a foundation on an active layer which does not swell is not any different from that which is employed in the areas that are free from permafrost. (Fig. 46). This is particularly true of structures which are so designed that they will have no thawing effect on permafrost. If, however, there is reason to expect thawing the strength of the ground when thawed and its ability to support the structure should be investigated and taken into account in the design of the structure.

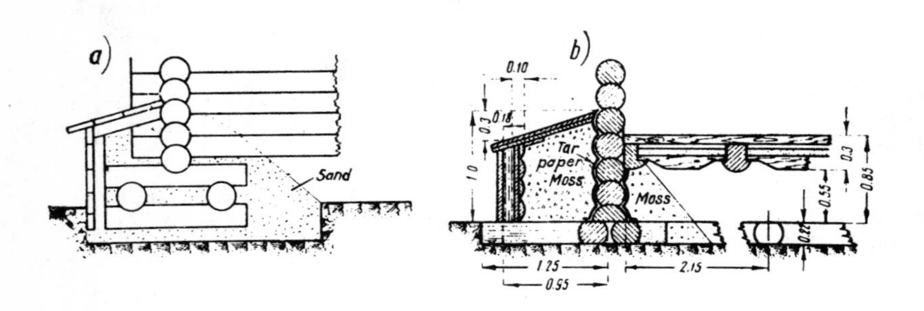

Simplified foundations for small wooden buildings.

(From Tsytovich and Sumgin)

In planning a foundation that will rest on or in the permafrost and where no thawing of ground is expected, the bearing strength of the underlying ground should be tested. If, however, thawing is expected a due allowance should be made for the consequent change in the strength of the ground.

The relative effect of the swelling of ground on a sunken foundation depends on:

1. Adfreezing strength of foundation to ground.
2. The depth of foundation.
3. Weight of the structure.

Observations show that the heaving of foundation begins when the adfreezing strength between the foundation and the frozen active layer exceeds the load plus the force of friction between the foundation and permafrost. With a foundation consisting of piles driven into the active layer the following process takes place during the swelling of ground. With the freezing of ground a pile is firmly gripped by the active layer and is heaved unless the load of the superstructure exceeds this force. When the ground thaws it commonly forms a plastic or fluid material (slud) which will tend to ooze out from beneath the foundation and cause the building to settle. The greatest damage to a building is caused by the unequal settling of the ground - the differential settling.

With piles which are driven through the active layer and a considerable depth into the permafrost the situation is different. Under this condition the frost-heaving will take place only if the force of heaving (a), (Fig. 47) (adfreezing strength of the active layer) is greater than the adfreezing strength of permafrost (b).

The types of foundation, regarded by Tsytovich and Sumgin as the most suitable with either passive or active method of construction, are those consisting of "footings", pillars, or piling. With the active method of construction the base of foundation should rest on ground below the lower limit of the active layer and the back-filled, non-frost ground around the foundation pillars should be well drained. An impervious wall should be constructed around the foundation to prevent infiltration of melt water from the active layer which surrounds the building. (Fig. 48).

Continuous foundations, particularly of masonry, are not satisfactory for the permafrost environment. Such foundations do not withstand the tensional strain, and because of considerable contact surface with the ground and a high heat conductivity coefficient, tend to disturb the thermal regime of the frozen ground.

The effect of a heated (inhabited) house on the thermal regime of the underlying ground is illustrated by the observations on the experimental house at the Petrovsk Permafrost Station in Transbaikalia. (Fig. 49).

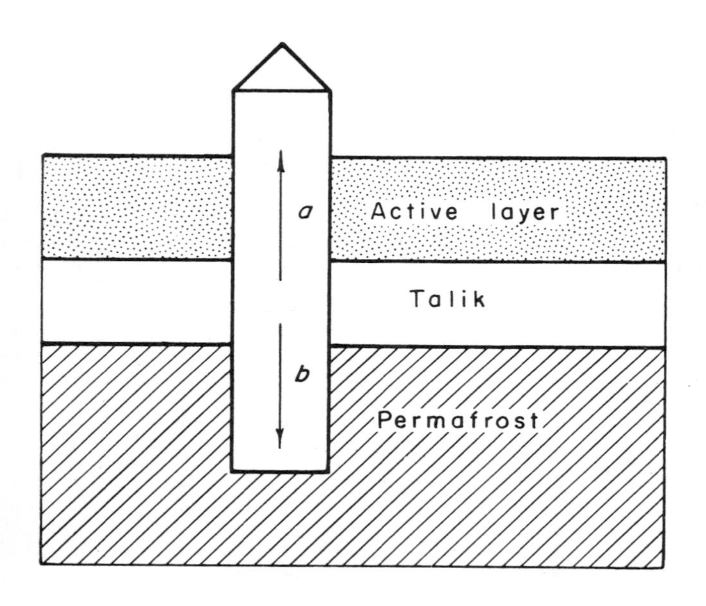

Diagram illustrating stresses in frost-heaving.

a-Heaving force of active layer.
b-Force of friction and adfreezing to permafrost.

(From Sumgin)

FIG. 47

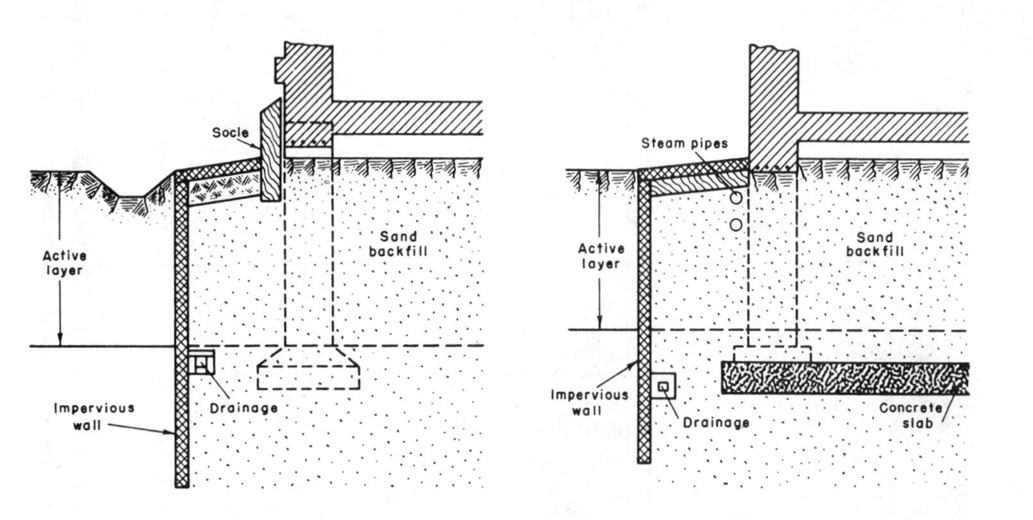

Types of foundations recommended in active method of construction – where permafrost is eliminated.

(After Tsytovich and Sumgin)

FIG. 48

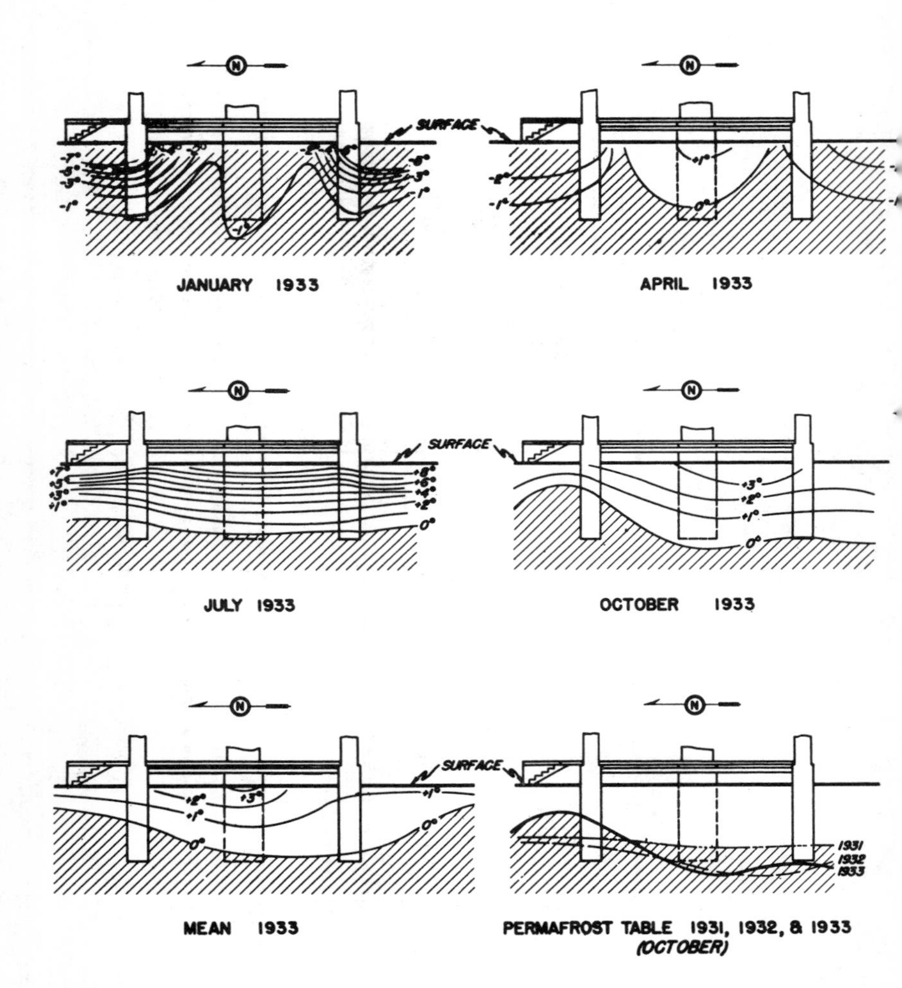

GROUND ISOTHERMS UNDER EXPERIMENTAL HOUSE IN TRANSBAIKAL

(HEATED HOUSE WITH MASONRY FOUNDATION. TEMPERATURES IN CENTIGRADE)

FIG. 49

According to Tsytovich and Sumgin, the study of the Petrovsk experimental house permits the following conclusions:

1. Houses with continuous masonry foundations noticeably disturb the thermal regime of the ground. However, the thawing of ground directly beneath the foundation is relatively slight as compared with the effect of the building as a whole.
2. The permafrost table shows an appreciable sag beneath and near the south-facing wall, whereas under the north-facing wall the permafrost table shows a definite rise. This condition is likely to result in a differential settling of the ground and a damage to the house.
3. There is a general lowering of the permafrost table underneath the building, as can be seen from the diagram "Mean isotherms of 1933". Should this trend continue the entire building may in time rest on thawed ground which may or may not be sufficiently stable to bear the load of the building.

In the permafrost area the ground on the south side of a building receives greater amount of solar radiation than on the north side. This causes a deeper thaw of the ground which produces an unequal settling of the building and cracking of walls.

The differential thawing of ground and subsequent settling of a building along the south-facing side can be appreciably minimized if the building is oriented with its longer dimension in the north-south direction. However, in the case of garages with sliding doors along one side of the building, it is more practical to orient the building with its longer axis in the east-west direction. The building so oriented may suffer some more or less uniform settling along the south or the north-facing wall without affecting the operation of the sliding doors. A long garage building oriented in a north-south direction and with sliding doors facing either east or west will have its doors jammed with the slightest differential settling along the south-facing wall.

In selecting the type of foundation the following frozen ground factors should be considered:

1. The structure and texture of the frozen ground.
2. The thickness of the permafrost and the behavior of the layer which overlies the permafrost.
3. Temperature regime of the ground.
4. Hydrologic regime of the ground.

A common cause of damage to buildings is the result of thawing of ground and especially of ground-ice beneath the foundation. No structures, therefore, should be erected on ground which has layers of ground-ice. This is particularly true of the buildings which are heated during the winter. Thawed ground with an excessive amount of moisture tends to become plastic, turns to slud, and is unable to support the weight of the building.

Swelling of ground, without available supply of additional water, measures only a few centimeters whereas when supply of water is available the swelling may be as much as 3-4 meters and occasionally attains a magnitude of tens of meters. Damage to a structure is also possible through horizontal stresses produced by the swelling of ground.

Piles or reinforced concrete foundations should be imbedded in permafrost to a sufficient depth to insure a firm anchorage capable of withstanding the uprooting force of the swelling ground.

To eliminate possible effects of horizontal stresses, the building should be completely surrounded by a trench or ditch which may be filled with gravel or other non-swelling material.

Structures which generate a considerable amount of heat (kilns, furnaces, etc.) should be constructed on ground that does not become weaker or less stable upon thawing. Such grounds are solid bedrock, fissured bedrock, if the fissures are free of ice or water, coarse sand and gravel. Sand and gravel will remain stable even if they are homogeneous.

It does not necessarily follow, however, that furnaces and similar structures, if appropriately designed, cannot be constructed on frozen ground of a different kind, without disturbing its permanently frozen condition.

For such construction two conditions should be taken into consideration:

1. When the active layer is subject to swelling a more rigid construction and firm anchoring of the foundation in the permafrost is required.

2. When the active layer is not saturated with moisture, is well drained and consequently is not subject to swelling, the foundation may rest upon the permafrost or be sunk into it only slightly and no provision need be made for the tensional stress which develops only where the ground swells.

In both cases appropriate measures should be taken to insure complete insulation.

Where the active layer swells, buildings should not be constructed on foundations which consist of flat and continuous supports resting on the surface, or of criss-crossed tiers of logs (cribbing). Foundations should be sunk and anchored in permanently frozen ground. The force of frost heave in uprooting pillar-like foundations is proportional to the perimeter of pillars, so the smaller the diameter of pillars the smaller will be the force of uprooting.

Flat footings fastened to the base of a concrete pillar or a pair of fastened horizontal cross-pieces - placed well within the zone of permafrost serve as effective anchoring devices. (Figs. 50 and 51)

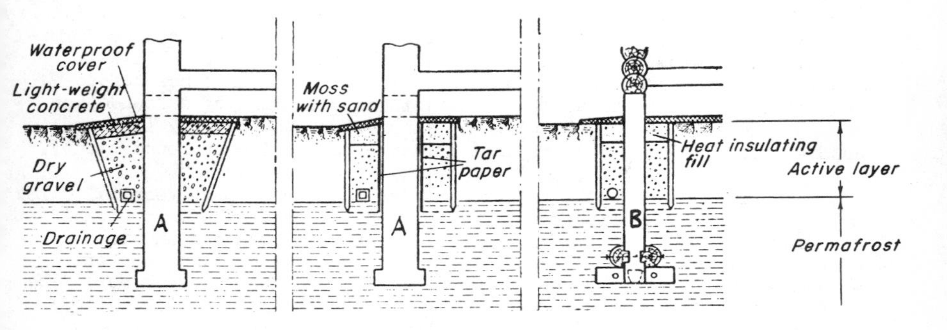

FIG. 50

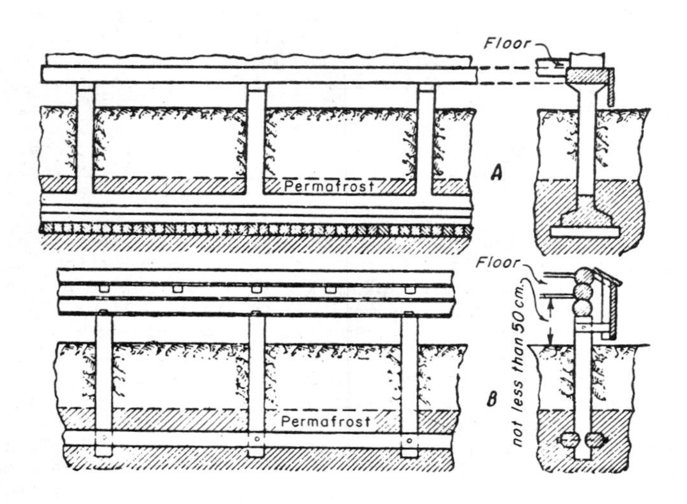

Anchoring of (A) concrete and (B) wooden foundations in permafrost

(From Bykov and Kapterev)

FIG. 51

For some structures it may not be necessary to sink the foundation into permafrost more than 25-30 cm. Ordinary buildings with floors above the ground tend to cause a rise in the permafrost table directly beneath the structure to as much as .8 to .9 meters. (Fig. 8). If a foundation is placed upon, or is only slightly imbedded in, the permafrost this subsequent rise in the level of the permafrost table will result in permanent freezing of the overlying ground that heretofore had not been frozen and will thus affect the anchoring of a foundation within a year. Care should be taken, therefore, to properly time such a construction. (See: Time element in the construction of foundations).

Anchoring may also be effected by deep penetration of vertical piles or pillars into the permafrost. Pillars buried or driven into permafrost to a depth of from 2.5 to 5 meters will provide a stable foundation for average buildings. It is generally safe to drive the piles into pre-thawed permafrost to the depth twice thickness of the active layer. The total length of the pile buried in the ground will thus be three times the thickness of the active layer. The only drawback to this type of construction is its high cost, but this is reduced if suitable mechanical devices for drilling of holes or pre-thawing of permafrost are employed. To insure a firm grip upon the pile by the permafrost, before the lowering of a pile into a hole, the hole is partly filled with thick mud which fills the space between the walls of the bore and the wooden or concrete pillar. After the freezing of the mud the pillar becomes securely anchored in permafrost.

Piles cannot be driven into permafrost by the usual methods. It is therefore necessary to resort to a preliminary thawing of the ground by steam points.

The part of the pile which is to be imbedded in permafrost should not be surfaced smooth and it is even advisable to have it notched or barbed. On the other hand the part that will be surrounded by the active layer should be made as smooth as possible and greased liberally with a water-repellent lubricant. Heavy cup grease or petrolatum with from 5% to 10% admixture of graphite may be used. Other lubricants should be tried to determine the longevity of their water-repellent property. To facilitate the handling of the greased piles and to insure the staying of the liberal coat of grease on the wood during the driving of the pile, the greased part of the pile may be wrapped with tar paper. (Figs 52 & 5 These measures make the anchoring of a pile in permafrost more secure and at the same time reduce the frictional hold on the pile by the active layer and minimize the pulling or uprooting force of the frost heave.

The tar paper should not be nailed to the pile as the nails and laths will rip the paper after the first winter freezing when the surrounding ground of the active layer will adfreeze to the outside of the paper and will heave it upward, sliding along the pile. The ripping of the tar paper will allow a more free access of outside moisture to the wood of the pile and will tend to deteriorate the water-repellent action of the lubricant. A sturdy cord tied around the pile to hold the tar paper in place should be tried.

Tapered ends of piles smoothed, greased, and wrapped in tar paper

FIG. 52

Piles driven into pre-thawed frozen ground. The hole on the left had been thawed more than is necessary.

FIG. 53

In general, care should be exercised not to rip the tar paper during the driving of the piles.

A section of the pile near the surface of the ground should be treated with creosote to prevent the rotting of wood and the disintegrating action of insects.

Deformation or damage to a building is frequently due to constructional defects, the commonest of which are:

1. Insufficient sinking of the foundation into permafrost which results in uprooting of the foundation.
2. Structures of considerable floor space with uneven heating and cooling of their different parts.
3. Foundations with overhanging sides or with uneven and rough surfaces facilitate heave by the active layer.
4. Continuous masonry foundations that have a large and uneven surface of contact with the ground.
5. Back-filling of foundation excavations with ground that swells.

These constructional defects account for the damage to a building only where the ground has a more or less normal amount of moisture. Occasionally foundations are built on ground that contains considerable seepage of ground-water, which may be under a very high hydrostatic pressure. In such cases the force of frost heaving may be so great that it is not possible to counteract it directly by any structural design. Therefore, foundations should not be constructed in ground with percolating ground-water but should be sunk as deeply as possible into the permafrost.

The following specifications should be applied to a building designed for ordinary living purposes and constructed by the passive method - retaining the normal thermal regime of the permafrost.

1. The distance from the surface of the ground to the floor joist should not be less than .5 or 1 meter.

2. Floor insulation should be such that the mean annual temperature in the air space below the floor is not allowed to rise above 1° C. (Fig. 54).

3. There should be a skirting (socle) (Fig. 55) surrounding the base of the building and forming the walls of the inter-pillar spaces. This requires no rigid specifications and may consist of thin boards (a and a' in Fig. 55) with the space between them filled with slag, dry dirt or any other insulating material. The socle should not rest directly on the ground but should have a clearance of from 10 to 15 cm. to prevent damage due to the swelling of ground. This gap, however, can be shut from the inside by leaning a plank against it and by piling some dirt or sand behind it. (Not shown in Fig.55).

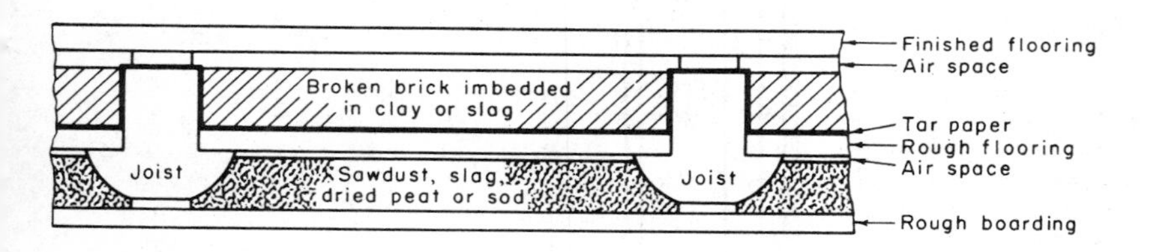

Design of floor recommended for living quarters in permafrost area.

(From Bykov)

FIG. 54

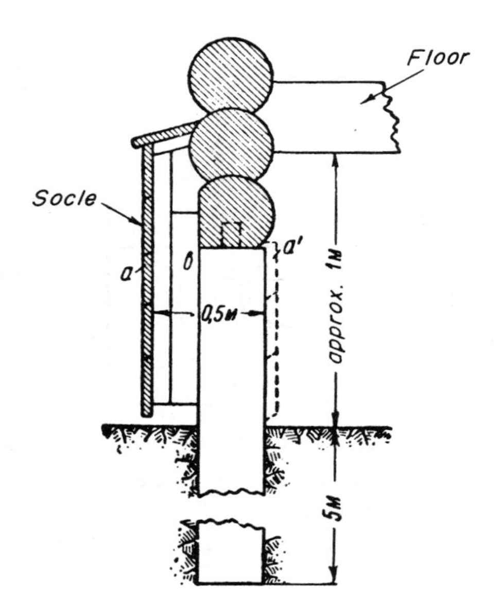

Design of Socle
(From Bykov and Kapterev)

FIG. 55

4. To maintain the thermal regime of the ground underneath the building the air vents in the socle should be shut in summer and open in winter. However, ventilation of the space between the floor and the ground becomes absolutely necessary if excessive dampness should develop during the summer. Bykov reports cases where shutting of air vents in the summer caused condensation of vapor to such an extent that water dripped from the floor joist.

Because of lack of necessary information no recommendations are offered in connection with the laying of steam pipes, and the installation of other utilities in conjunction with the construction of building foundations. It is hoped that these problems will be solved in the near future though:

1. Careful study of insulating property of various structural materials under variable conditions of temperature and moisture.
2. Study of heat conductivity of different structural materials under the conditions of permafrost and active layer, with variable conditions of temperature and moisture.
3. Design of boilers, motors and other heat generating installations to allow easy and effective insulation.

In summary, the measures to counteract frost heaving of foundations are:

1. To prevent or minimize the freezing of the active layer to the foundation:
   Construction of muffs or loose collars around the foundations and piles has been shown to be effective for one year and rarely for two years. The technique is not yet sufficiently perfected to insure permanent safety against frost heaving of foundations. The muffs or collars appear to work smoothly during the first year after installation but adfreeze to the foundation the following year. (Fig. 56).
2. To remove the causes of swelling by draining the ground or by replacing it with non-swelling ground:
   Drainage of the ground is practicable only under conditions of sufficient relief. In the undrained ground filling of the foundation pits with gravel, slag, or coarse sand and surrounding pillars by the same materials is useless and even detrimental.
3. To cover the surface of the foundation with substances that will prevent or greatly minimize adfreezing.
   Experiments have been made with parafin to reduce the cohesion between the frozen ground of the active layer and the pillar. Some of these parafin-saturated and coated pillars were anchored by additional weight but as yet no positive results have been achieved.
4. To anchor or add weight to a foundation to counteract the uprooting force of frost heaving.

Anchoring of foundations brings satisfactory results only by anchoring in permafrost.
Unheated and uninhabited buildings may be constructed resting on logs placed upon the surface. They will undergo some frost heaving and settling but the effect will not be cumulative.
When an unheated building such as a warehouse is placed on a sunken foundation it is usually uneconomical to sink the foundation to a safe depth and as result the building suffers maximum effect of frost heaving and, owing to the annual increase of residual swelling, the building may become seriously damaged.

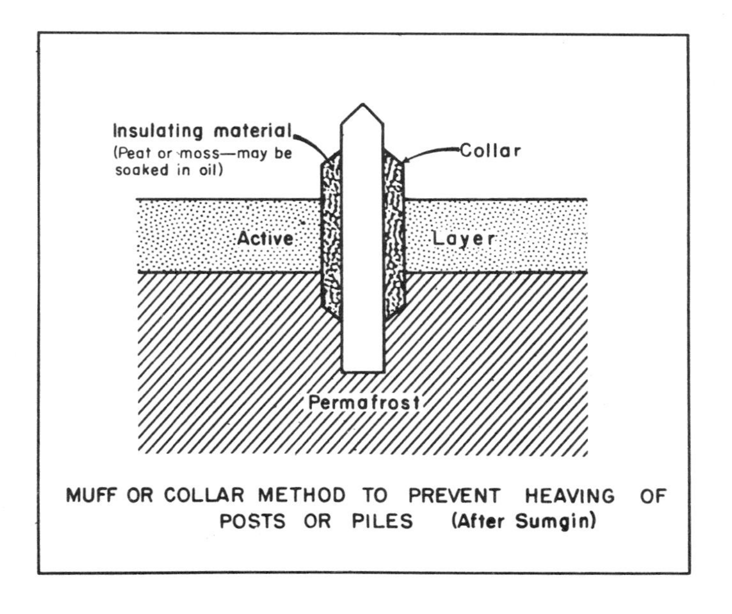

MUFF OR COLLAR METHOD TO PREVENT HEAVING OF POSTS OR PILES (After Sumgin)

FIG. 56

When, in spite of preventive steps, structures begin to deform, thermal insulation may be resorted to as a corrective measure.

Thermal insulation may be of stationary or portable type. Stationary thermal insulation consists of a layer of material having low heat conductivity such as peat, moss, slag, or of specially manufactured products such as Celotex, asbestos, etc. This insulating material is laid down in a strip 1.5 to 2 meters wide around the foundation.

Portable insulation consists of sheets of pressed peat, moss, or other insulating material about 1 m. x 1.5 m. which may be carried from place to place and which are laid down around a foundation.

The portable insulation method is most satisfactory in attaining the desired control of thermal conditions. Its chief disadvantages are greater expense and need of trained personnel.

The stationary insulation is less satisfactory as the material used is likely to absorb moisture which will reduce its heat conductivity.

The time of installing stationary insulation will depend on the following considerations: If a foundation is built on the plan of retaining the permafrost conditions (passive method) the insulating layer should be laid down during the summer and dry. If, on the other hand, it is planned to prevent deep penetration of ground frost (active method), insulation should be installed at the time of maximum freezing in January or February.

Time Element in the Construction of Foundations. Driving piles into the pre-thawed ground should be so timed that there will be no chance of uprooting the piles by the swelling of ground before they are firmly anchored. The best time for driving piles is between February and June. In February the swelling of the previous winter begins to disappear and does not set in again until about November. Should it become impossible to so time the work the ground of the active layer should be removed from around the pile to the depth of normal seasonal freezing and the pile, after being liberally greased, covered with tar paper. After that the ground of the active layer may again be filled around the pile. Grease should not be applied to piles before they are driven into ground as there is always a chance that some of it may accidentally get on that part of the pile that will penetrate permafrost thus preventing proper anchoring.

Experimental work shows that at a depth of 5 or 6 meters below the permafrost table the freezing of a pile to permafrost takes about a month and a half to 2 months. Only after that time can the pile be loaded with the calculated weight of the projected structure.

After driving the piles according to a pre-thawed method, thermal balance in permafrost is usually completely restored in about 6 or 7 months, provided the piles are not too closely spaced and do not occupy a considerable area of the ground.

As is the case in driving piles, it is important that a foundation designed to be anchored in permafrost should be constructed during the most favorable time between February and November so as to allow time for the freezing of the "anchor" before swelling forces of the following winter develop. This consideration, however, does not hold if the vertical load upon the foundation units is known to be greater than the heaving force of the swelling ground.

Anchoring may also be accomplished by building a foundation immediately after the swelling effect of the ground has occurred, say from December to February. After the foundation is laid the excavation is then filled in with frozen ground. With the advent of summer, moisture from the active layer will seep into this porous fill and freeze it. Unfrozen fill may also be used if it is tamped down and allowed to freeze through.

Foundations that are built in the spring (April to early June) and filled in with unfrozen ground will adfreeze to the permafrost within a normal period of time provided the volume of the fill is not too great.

Foundations constructed prior to the period of ground swelling may heave slightly. If such a heave is likely to cause serious damage it is then necessary to adopt the measures recommended for setting piles. The part of the foundation that will normally be in the active layer should be left exposed (not covered with backfill) until the lower part of the foundation is firmly anchored by adfreezing to the permafrost. An alternative measure consists of the application of grease and then covering with tar paper before the active layer zone of the excavation is filled in with dirt.

Replacement of Ground Susceptible to Swelling. Where the ground is composed of clayey or silty material it is almost impossible to eliminate entirely the destructive effect of frost swelling. Draining such ground may materially decrease the amount of moisture in it, but there will always be enough moisture left behind to cause appreciable swelling during the winter freezing. Under these conditions it is advisable to replace the undesirable ground with such non-swelling materials as clean gravel or coarse sand.

Measures should, of course be taken to prevent subsequent silting of gravel or of sand. Subsequent silting may be prevented by building around the gravelly fill a vertical wall of two-inch boards parged on the outside with 4 inches to a foot of impervious clay. The board wall should penetrate the permafrost at least half a meter. The upper edge of the clay parge on the board wall should connect with the horizontal top layer which prevents the infiltration of surface waters[4].

4/ Lukashev casts some doubt on the efficacy of this protective layer of clay, for in his opinion, although a layer of clay a foot or so thick may prevent silting, the clay itself will be a contributing factor to greater swelling.

In view of the fact that the gravelly and bouldery fill have a higher heat conductivity than silty or clayey ground, it is necessary to combine this insulation against silting with thermal insulation in order that the underlying permafrost may not thaw and disturb the thermal balance.

The depth to which the unsuitable ground should be removed and filled in with the gravelly material must be determined by the thickness of the active layer.

Precautions Against Damage by Settling and Caving. Measures to safeguard structures from damage due to settling and caving are:

1. Construction should not be undertaken if it will disturb the minor relief features adjacent to the structure and will cause lowering of permafrost table.
2. The thick natural cover of grass, moss, peat, and shrubs should be maintained.
3. Structures should be painted white.
4. Foundation should rest on a raft-like platform of logs, excavations should be lined with boards or other insulating cover and the space between the floor and the ground should not be "dead air" but should be ventilated.
5. The load on a foundation resting on the unfrozen ground should be computed with a wide margin of safety.
6. Construction of a substantial plate-like basal part of the foundation ("shoe") of concrete capable of withstanding a considerable torsional stress from uneven thawing of the underlying ground.
7. The weight of the structure should be centered with the center of gravity of the base of foundation.

The least settling takes place where the bottom of foundation is separated from the permafrost by a layer of unfrozen sand. Where this layer of sand is of sufficient thickness the pressure exerted by foundation is distributed over a considerably greater area. This minimizes if not entirely eliminates the effect of uneven settling or caving. (Fig. 57).

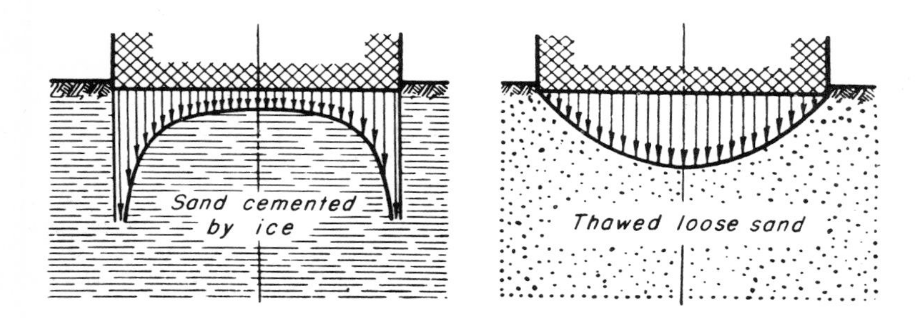

Distribution of pressure beneath a rigid foundation.
(From Tsytovich and Sumgin)

FIG. 57

Lenses of ice and plastic or semifluid ground (slud) are frequently encountered during excavation. In order to avoid strong settling and caving, the ice and the unstable plastic ground should be removed and backfilled with a non-swelling material such as sand or gravel.

When stability of a structure can be attained only through a radical alteration of existing hydrologic and permafrost conditions, it is then necessary to:

1. Regulate the hydrologic regime of the surrounding area.
2. Construct supporting walls and drive reinforcing piles.

These measures can be successfully carried out only if they are based on a thorough knowledge of hydrologic and permafrost conditions of the locality.

Drainage. Areas with excessive moisture should be drained and seepages checked.

Where the permafrost table is fairly close to the surface the following factors should be considered.

1. Composition of the upper part of permafrost.
2. Whether the active or passive method of construction is to be employed.

Where permafrost is composed of silt and fine sand with a considerable proportion of ice, drainage will not give satisfactory results because the thawed ground will become plastic or semifluid (slud). Under such conditions drainage ditches will slump and erode away, eating into the surrounding ground and endangering newly erected structures. (Fig. 58).

Apparently there is no completely satisfactory method of drainage under the conditions just described. The critical time, when drainage should function properly, is during the spring. Current practice of covering the ditches with insulating layers of peat and moss appears to be unsatisfactory because this cover delays the thawing of ditches in the spring when they are most needed. It is recommended that in the autumn when drainage is not required, it should be stopped and speeded up in the spring.

Chernishev recommends filling of drainage ditches with material of higher heat conductivity, such as boulders, gravels, or coarse sand, capped with a layer of clay to prevent silting.

Drainage measures should be carried out considerably in advance of building as there is always a possibility of some unexpected unfavorable effect on the ground that has been selected as the site.

Excessive moisture of the ground may be due to inflow of surface waters or to underground seepages.

Elimination of surface water does not present any serious problems, but water percolating underground is more difficult to control. It has to be intercepted and diverted into deep wells or sumps. Sinking of wells requires particularly thorough investigation of the hydrologic conditions. Existence of a strong seepage from below the permafrost may result in a rapid filling of the well and even an over-flow which will make the conditions considerably worse.

Solifluction and slumping caused by deep thawing (and flowing) of ground in a poorly constructed drainage ditch.

(From Lukashev.)

FIG. 58

Road Building

The choice of route in the permafrost area should not be dictated by the usual evaluation of topography or by the premise that the shortest and the most direct route is the most satisfactory. The success of construction and particularly of maintenance of a road depends to a great extent on the location of the roadway with reference to the slopes of the relief, the angle of these slopes, their thermal and hydraulic regimes, and the nature of the ground. The route should be changed, even though the distance is increased, if a preliminary survey shows that the original route traverses an unsatisfactory terrain. Availability and easy access to satisfactory road metal also should be given careful consideration.

Swelling, settling and caving of ground due to the melting of ground-ice, landslides and icings are the major causes of damage.

The following conditions in the subgrade or near the road usually cause swelling and consequent heaving of the surface:

1. A layer or a small mass of fine-grained (silty) material enclosed in a coarser and better drained material.
2. Clays containing inclusions of fine sand or silt.
3. Medium and fine grained sands saturated with water which cannot seep down because of an underlying impervious layer of clay or permafrost.
4. Coarse textured ground containing water under hydrostatic pressure.

It is important to remember that the permafrost table will rise beneath the fill and will even extend into it if the height of the fill is in excess of the thickness of the active layer. Also seasonal thawing will penetrate somewhat deeper into a fill than in the surrounding ground, especially on its south-facing side. (Fig. 59). The deeper thawing on the south side of a fill commonly produces a slight depression in the permafrost table just below the south-facing shoulder. Excessive moisture may accumulate in this depression and cause a landslide. This may be prevented by surfacing the south slope of the fill with a layer of some material with low heat conductivity, such as: sod, peat and, wherever available, slag. Peat, being inflamable, should be covered with sod. A protective layer of sod is also recommended for non-inflamable porous materials to reduce the convection of warm air.

Particular care should be exercised in laying out roads in swamps and in areas where underground ice is known to exist. Drainage operations and construction of fills in such areas should be carried on with caution. Swamps are frequently underlain at a moderate depth by either permafrost or ice.

Drainage ditches and canals are subject to active erosion in depth as well as in width, causing a lowering of the level of permafrost. Excessive erosion of drainage ditches with gently sloping sides may be checked by lining the bottom and sides of the ditch with sod or peat, provided the flow of the water is not very strong. Ditches lined with burlap in a shingled arrangement soaked in tar or oil, have also proved to be effective in checking excessive erosion of ditches. Wide and shallow drainage ditches are, as a rule, less subject to erosion and slumping. Removal of the vegetative cover may also cause a considerable lowering of the permafrost table and the newly thawed oversaturated ground turns into a slud which cannot support the weight of the fill.

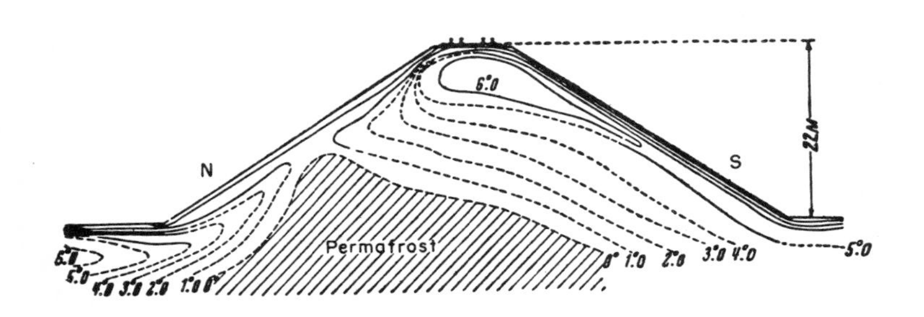

High road fill in September. Ground isotherms show deeper thaw of ground on the south side than on the north side.

(From Sumgin)

FIG. 59

The hydrologic regime of the area traversed by a newly constructed road fill may also be affected by the fill. This effect may be very slight if the direction of the fill more or less coincides with that of the flow of surface and ground-waters. On the other hand if the fill runs at right angle to the direction of flow of water the effect may be quite marked and may lead to serious consequences. Such a situation is illustrated in Fig.60 .

Here the swampy condition is intensified on the up-hill side whereas on the down-hill side the swamp is more or less thoroughly drained. The drying of the swamp on the down-hill side causes the lowering of the permafrost table and thawing of ground previously frozen and firm. This oversaturated ground becomes plastic or even fluid on thawing and seriousl

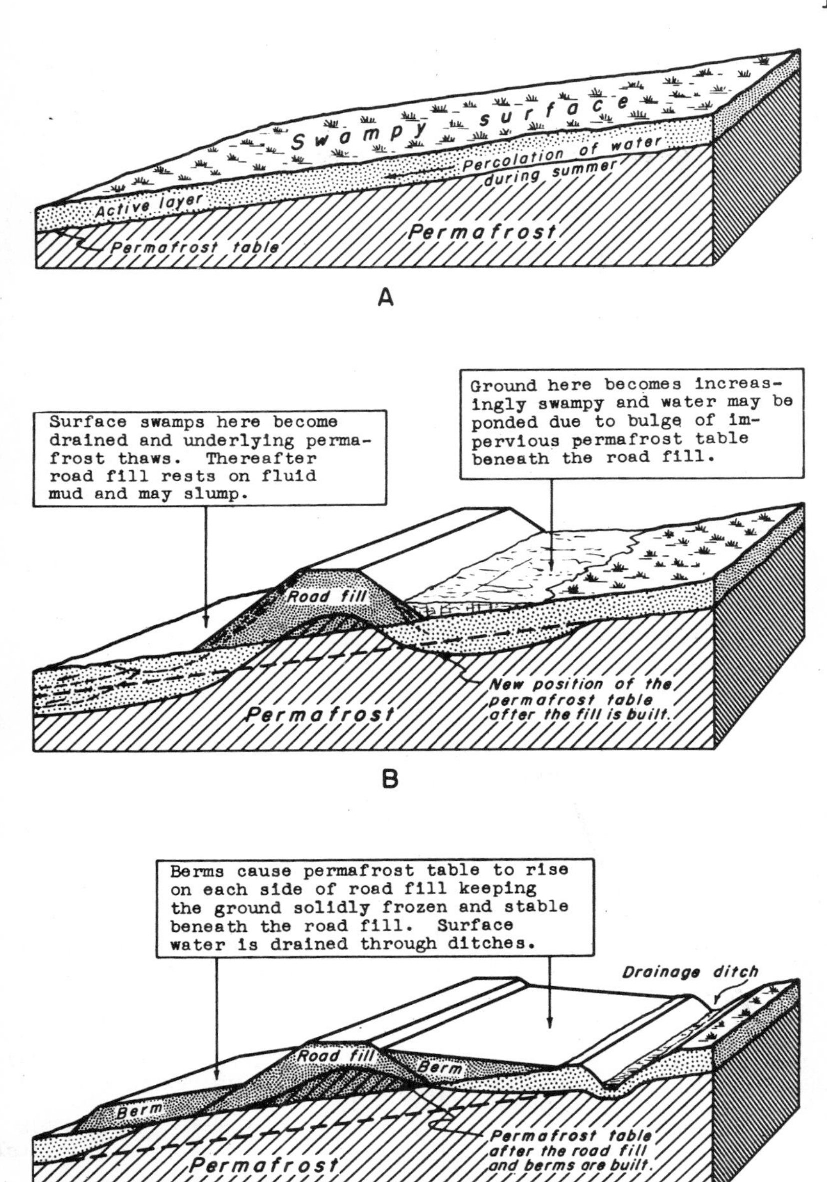
Swampy surface
Percolation of water during summer
Active layer
Permafrost
Permafrost table
A
Surface swamps here become drained and underlying permafrost thaws. Thereafter road fill rests on fluid mud and may slump.
Ground here becomes increasingly swampy and water may be ponded due to bulge of impervious permafrost table beneath the road fill.
Road fill
New position of the permafrost table after the fill is built.
Permafrost
B
Berms cause permafrost table to rise on each side of road fill keeping the ground solidly frozen and stable beneath the road fill. Surface water is drained through ditches.
Drainage ditch
Road fill
Berm
Berm
Permafrost table after the road fill and berms are built.
Permafrost
C

FIG. 60

impairs the stability of the fill. Water which accumulates on the up-hill side of the fill will start to move along the fill and will cause some sagging in the permafrost table directly under it, while the permafrost table beneath the fill will rise slightly. This situation will be favorable for landslides and subsequent damage to the fill. A drainage canal will prevent the accumulation of water above the fill. Protective benches (berms) on both sides of the fill will check the lowering of the permafrost table near the fill. These berms should be made of material that is a poor conductor of heat. Properly located culverts will also eliminate some of the undesirable conditions described.

The effect of berms in preserving the frozen condition beneath the fill, and particularly in preventing the thawing of underground ice, is illustrated in Fig. 61 .

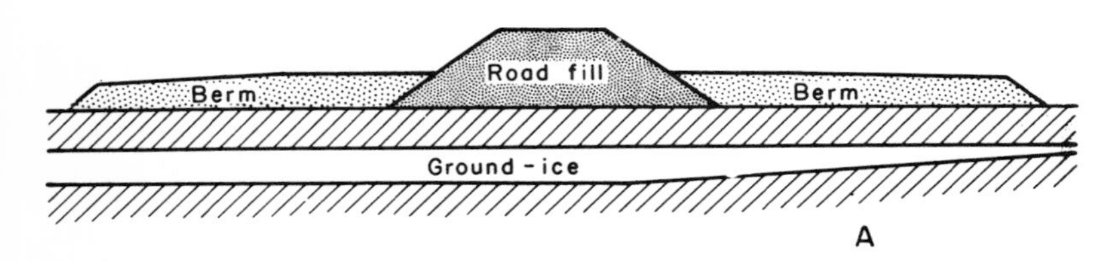

Melting of ground-ice and slumping of road fill is checked by the construction of berms.

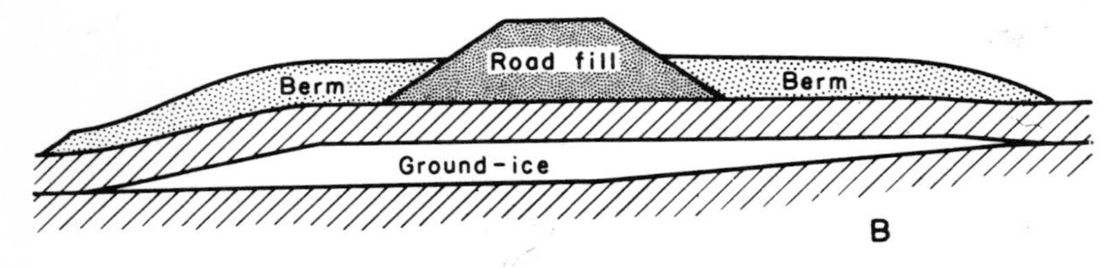

(After Bykov and Kapterev)

Effect of berms in preserving the frozen condition beneath a road fill.

FIG. 61

Road cuts present even greater difficulties. The ground of the active layer is usually either partly or completely removed and frequently even the permafrost is penetrated. This causes a substantial lowering of the permafrost table and the ground that was formerly frozen and firm is thawed and changes into a less stable plastic or freely flowing material which gives rise to numerous landslides and slumps. The only solution of this problem appears to be a replacement of the unstable ground by more suitable material, which is a very costly undertaking.

Another serious difficulty in the construction and maintenance of roads is caused by icings or, as they are sometimes called, "ice-fields". A seepage of ground-water or of river water from beneath the ice may freeze on or near a road in successive sheets of ice that may attain a thickness of several meters. In Siberia, on the new Amur-Yakutsk Highway over one hundred large icings have been recorded within a distance of about 700 kilometers. Stretches of road each several hundred meters in length have been rendered entirely impassable due to these icings.

Behavior of an icing and its destructive effect on a road is illustrated by the Onon River icing on the Amur-Yakutsk Highway, 124 km. north of Skovorodino, Siberia. (Fig.62).

The Onon River valley, where it is crossed by the highway, is about 250 meters wide. The bottom of the valley is bordered on both sides by steep slopes with small mountain brooks which flow into the Onon River. Because of their rapid flow these tributary brooks freeze much later in the winter than the Onon River.

Permafrost in the bottom of the valley begins at .6 meter below the surface but on the valley sides it is encountered at a depth of 1.5 meters. During the winter the percolating ground-water beneath the frozen surface ground becomes constricted in a narrow passage-way or is blocked and trapped where the active layer freezes clear through to the permafrost. Such an obstruction to the underground seepage of water was created along the sides of the highway where the surface had been scraped or trampled down permitting the ground underneath it to freeze to a greater depth much earlier than in the adjacent undisturbed ground.

Six icing-mounds formed along this part of the road. (Fig.62). The second mound from the south (indicated by dotted lines) after a preliminary cracking and trembling suddenly exploded with a loud sound resembling that of a cannon. Large slabs of ice 2 meters thick and up to 19 meters long were thrown out and carried 120 meters down the valley across the road (Fig.63) by a 75 meter wide rushing torrent. The water which poured out of the exploded mound spread down the valley for a distance of about 5 kilometers. A small highway bridge in the path of this torrent was shaved off to its foundation, shrubs were flattened down to the ground, and the bark on large trees was badly scarred by the moving pieces of ice. (Fig. 64)

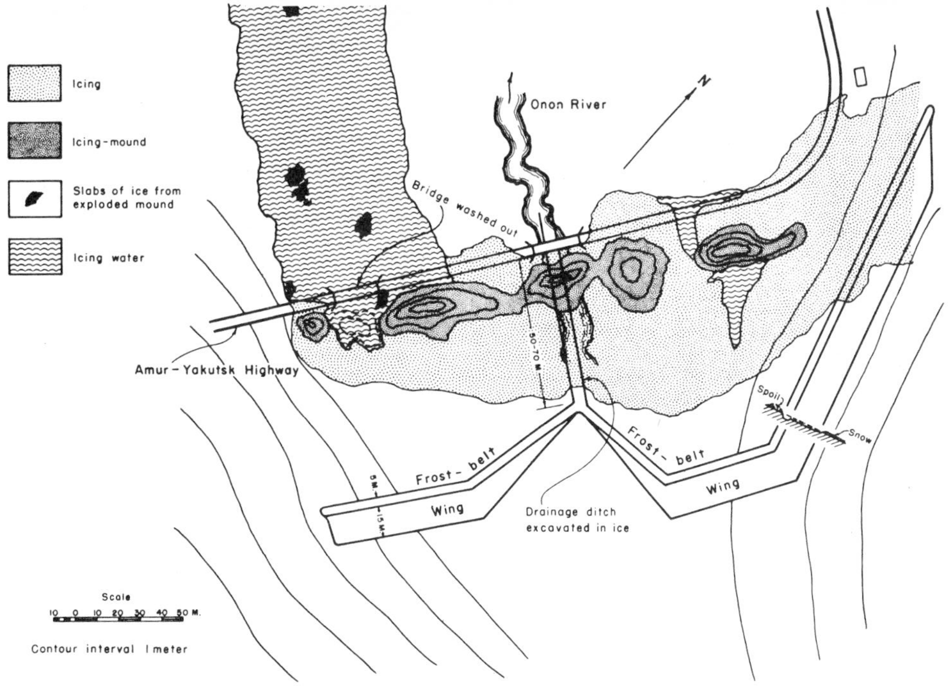

Plan of icing along the Amur-Yakutsk Highway at the Onon River crossing and plan of a projected FROST-BELT. In the upper left corner: Flooded area with slabs of ice as a result of the explosion of icing-mound shown by dotted lines. (After V. G. Petrov)

FIG. 62

Slab of ice 2 meters thick and 7 meters long was carried 120 meters down the Onon River valley following the explosion of an icing-mound. Slabs in the distance are 19 and 11 meters long. (From V.G. Petrov.)

FIG. 63

Remnants of icing-mound after the explosion. (From V.G. Petrov.)

FIG. 64

This catastrophic action lasted only a couple of hours. Prior to the explosion, the sixth icing-mound, about 150 meters to the northeast was also showing signs of unrest but the explosion of the second mound relieved the underground hydrostatic pressure and caused the sixth mound to subside and to cease cracking.

The Onon River icing, described above, clearly demonstrates the importance of hydrostatic pressure in the formation of mounds and in their destructive effect on man-made structures.

In Siberia hundreds of river and ground icings along roads showed a remarkable alinement with the roadbeds, following every curve of the road, as shown in Fig. 65.

A satisfactory preventive or corrective measure against icings, known as the frost-belt, has been developed by V. G. Petrov. It consists of a specially constructed ditch the primary object of which is not so much to drain the water as to cause an early and complete freezing of the active layer far enough from the road so that icing will not damage the roadbed. (Figs. 66 and 67).

The action of the frost-belt is illustrated in Fig.68 where the icing "A" was eliminated by causing the water that fed it to form another icing within the confines of the frost-belt some distance up-slope and away from the road. Vegetation was removed and a dtich about 1 meter deep dug at "m" some distance from the roadbed "B-B". Subsequent freezing of the ground above the dashed line "C" will be most intense along this ditch. The freezing will reach the level of permafrost and will prevent the underground percolation of water below the ditch into the area near the road. As a result an icing will form either in the dtich or just above it, as is indicated by "A".

The frost-belt should be 5 to 10 meters wide and between 0.5 and 1 meter deep. It can be 1 meter deep and only 5 meters wide if it is bordered on the up-slope side by an additional strip from 10 to 15 meters wide which is stripped of vegetation and sod. In such a case the profile of the frost-belt will look like an outline of a dipper. The frost-belt should be located from 50 to 100 meters away from the road and should be as long or slightly longer than the icing which is to be eradicated. In a construction of the frost-belt ditch the spoil should be piled on the down-slope side and if any snow is present it should be scraped up-slope and left lying immediately adjacent to the upper edge of the belt.

If large icing-mounds form above the frost-belt they should be punctured and drained before they explode and cause damage. They may be punctured either by simply chopping the ice, by blasting, or by melting the ice with thermite.

To be effective a frost-belt should be constructed early in the winter before the first snowfall, preferably before the beginning of

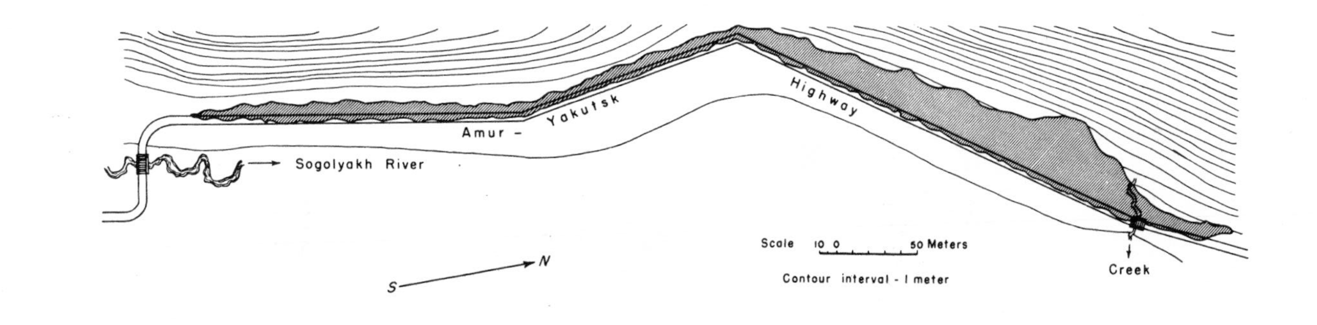

Plan of icing along the Amur-Yakutsk Highway, 91 km. north of Skovorodino
(Surveyed March 16, 1928)

(From V. G. Petrov)

FIG. 65

Excavation of FROST-BELT ditch along the Amur-Yakutsk Highway, 169 km. north of Skovorodino, Siberia. Photographed in November 1929. (From V.G. Petrov)

FIG. 66

The same FROST-BELT in operation on March 20th, 1930. The ditch is filled with icing, icing-mounds, and icing water. (From V.G. Petrov)

FIG. 67

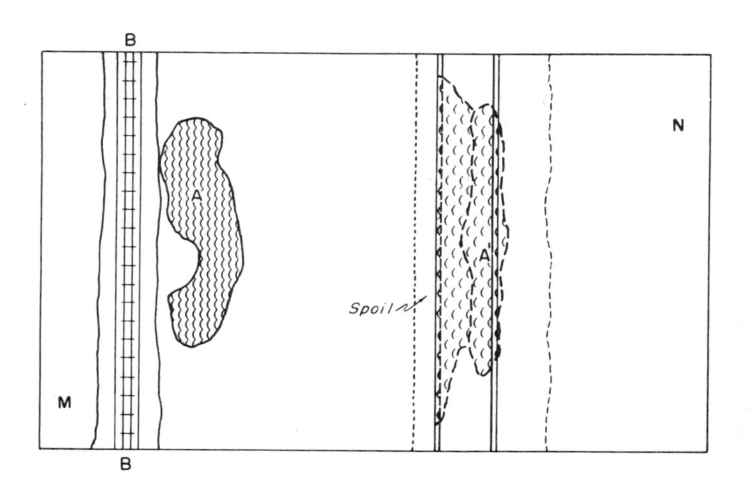

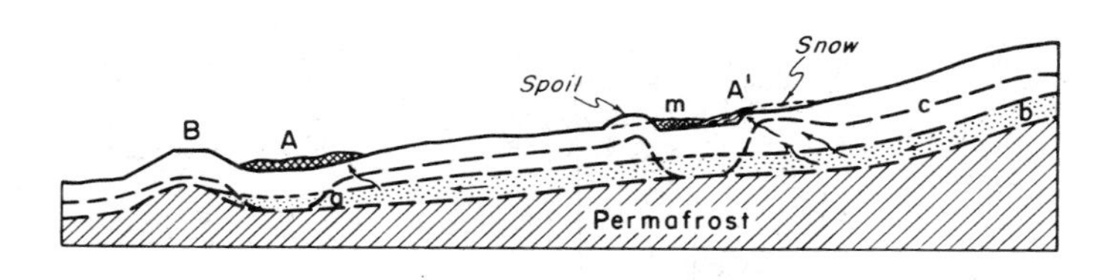

Elimination of icing (A) by FROST-BELT METHOD

(After Petrov)

FIG. 68

freezing weather. The snow should be kept off the frost-belt until the end of January. After that time it may be left lying on the frost-belt as it will tend to preserve the low temperature of the ground which is just what is desired.

In places where the ground is free of natural insulating material such as sod, moss, or peat, the frost-belt ditch is not needed. Freezing of ground along the belt can be achieved by simply removing the snow.

The frost-belt method gives satisfactory results only if the icing is fed by the surficial water. The method is likely to be less effective or may fail altogether if the icing water comes from a deep source. According to Yanovsky, icings that are fed by deep waters can be better controlled by diversion ditches.

As the source of water that forms the icing is not always easily determined the planning of preventive measures should be always preceded by a thorough study of the icing as outlined in the "Survey of the icing".

The inadequacy of the frost-belt method of Petrov is that after a year or two of successful operation, the level of the permafrost table will develop a sag corresponding to the surface profile of the ditch and the percolation of water will continue unchecked to the fill of the road-bed. To avoid this Tsytovich and Sumgin recommend filling in the ditch or covering it with some insulating material in spring and opening it again in the fall. Such a maintenance of the frost-belt ditch is undoubtedly effective but is costly.

A variant of frost-belt proposed by Bykov and Kapterev appears to be most satisfactory. (Fig. 69). Instead of the ditch "m" a trench of 1 to 1.5 meters wide is cut across the water-bearing layer to a depth of about 2-2.5 meters. This trench is then filled with clay or with water-saturated silty or clayey ground, and is well tamped down. A row of pile-planks is then driven into this ground and both sides of the exposed planks are covered with fill. This fill will cause the permafrost table to rise and as a result the planks will be firmly anchored.

The water percolating down the slope will be forced to the surface by the frozen ground and by the impervious wall of the trench and will form an icing on the up-hill side of the barrier.

A slightly different procedure is followed in building a frost-belt to offset a river icing. One type, used when the road crosses a shallow river which freezes completely in winter, consists of a ditch paralleling the road and 200-300 meters up-stream from the road. The ditch in the river ice should be 3-4 meters wide and about 2 meters deep. The excavation of ice is usually started just as soon as the river ice is thick enough for operation. Excavation is continued gradually until the bottom of the river is reached. The ends of the ditch are excavated in both banks of the stream and are usually constructed during the summer.

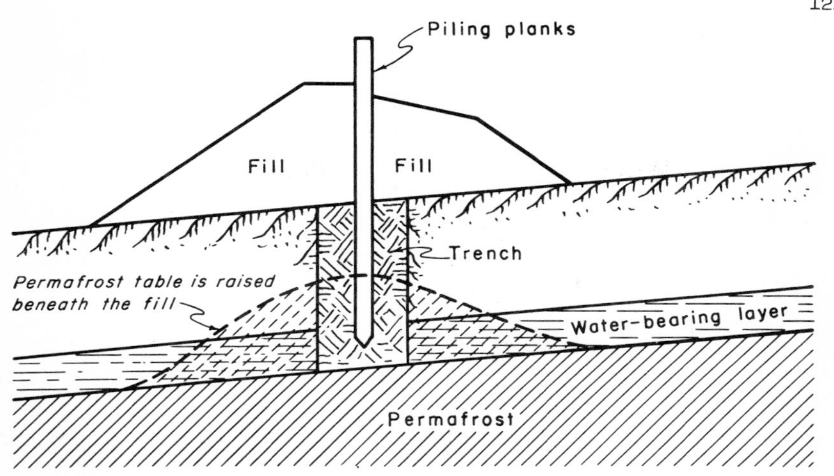

A variant of FROST-BELT proposed by Bykov and Kapterev.

(From Bykov and Kapterev)

FIG. 69

The spoil (ice as well as ground) is piled on the down-stream side of the ditch to form a dike to hold the icing water and to prevent its spreading onto the road. The up-stream side of the ditch should rise to the surface at a gentle angle.

Should an excessive amount of water accumulate in a frost-belt, a drainage ditch cut in the ice down stream and under the bridge may become necessary.

Cribbing set in the river bottom some distance above the road may also act as a frost-belt and shift the formation of icing up-stream away from the road.

Destructive river icing may also be averted by building snow banks on ice across the river at shallow places and near rapids above the road. At the same time a hole should be chopped in the ice below the road to drain the water and thus to remove the possibility of creating hydrostatic pressure.

A mild river icing may be diverted from a road by sufficiently high earth-filled shoulders (levees).

Gravel surfaced roads are frequently made impassable due to the presence of frost-boils. Frost-boils are moisture saturated, semi-fluid pockets of ground which develop during the spring thaw and which when broken through by the heavy traffic form a regular quagmire. (Fig.70).

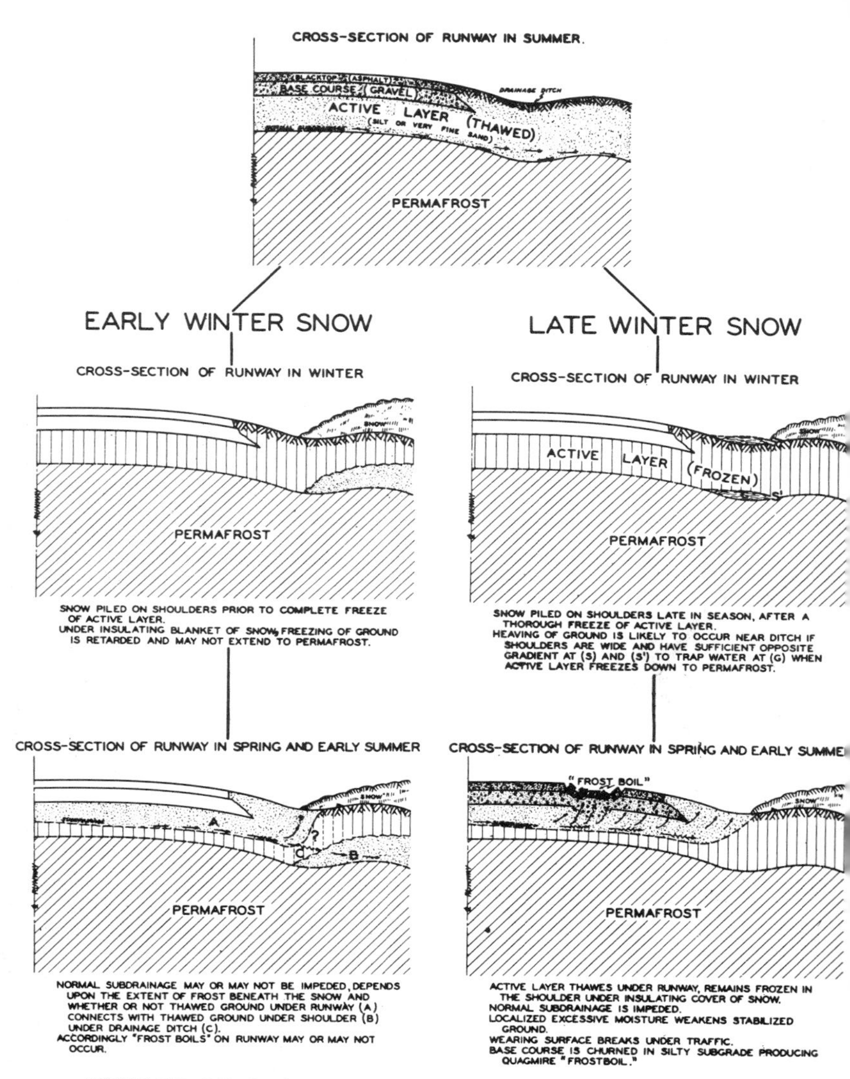

EFFECT OF SNOW PILED ON SHOULDERS OF RUNWAY

FIG. 70

Modern highway practice of keeping the roads clear of snow produces a deeper penetration of frost beneath the travelled road than under the shoulders and side slopes and in the spring the surface of the road begins to thaw first and proceeds at a more rapid rate than under the shoulder where the ground is usually protected by either sod, grass, or snow. The thawing of ice layers beneath the roadway produces an excessive amount of water which is unable to escape through the still frozen ground below and so remains along or near the center of roadway. It may "boil" up through small crevices in the surface of the road or may remain undetected beneath the surfacing. The surface of the road may be dry and dusty but ridges and ruts may begin to develop because the underlying ground behaves as a semifluid mass. When the crust is broken by traffic the gravel will be churned with mud. Many cures for frost-boils have been employed in road engineering but most are unsuccessful.

Simply dumping of gravel or rocks into an open boil is as fruitful as dumping them into the sea.

Melting the underlying frozen subgrade by steam to obtain vertical drainage has met only with partial success. Subgrade ground even when thawed may be very impervious.

Installation of tile drains also does not always remedy the situation where the subgrade fill has a high capillary lift.

The most satisfactory cure for boils, described by Motl, (Fig. 71), consists of a trench along the center of the roadway with a perforated pipe and a backfill of gravel. According to Motl, metal pipe is preferred to tile because it is better able to withstand the differential deformation during the winter swelling of ground. Frost boils may also be eliminated if the entire width of the roadway is excavated and filled in with at least 30 inches of gravel and rock and two lines of drains are laid along the edges of this fill.

Bridges

Construction of bridge piers is essentially similar to that of foundations of other large structures. It should be remembered, however, that the permafrost table is likely to have a prominent depression under the stream bed due to the thermal effect of flowing water. Therefore the principles of construction have to be based on the rules which apply to conditions of active layer and not to those of the permafrost ground.

Proper precautionary measures should be taken to eliminate possible damage from river icing during the winter. Pier foundations should be imbedded in permafrost ground to a depth of about 1 meter and placed on a wooden platform built of criss-crossed logs (20 cm. in diameter). This platform should extend at least 20 cm. beyond the perimeter of the pier. These "cushions" serve to stabilize the piers as well as to permit normal setting of cement, which may freeze if poured directly upon the frozen ground. This method of "cushions" is necessary only where thawing of the permafrost will make the superstructure unstable. If the ground consists

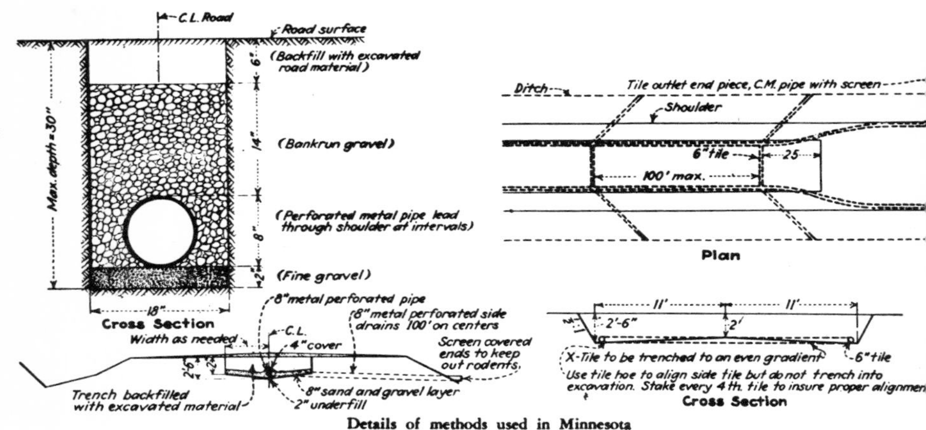

Details of methods used in Minnesota

Left, above—Narrow trench method. Left, below—Wide intercepting gravel layer and perforated pipe. Right—Massive type of tile drain for use under pavements.

FIG. 71

of solid bedrock cushions are not necessary. The use of joined piles to support the bridge is not recommended. Joined piles, anchored in permafrost, tend to break at the join due to the frost heaving of the active layer.

Bridge piers located where the river is fairly deep are less likely to be damaged by freezing than where the stream is shallow. Shallow stretches of rivers usually freeze more quickly and to a considerably greater depth. Freezing in a shallow place may penetrate to the very bottom of the stream and may even extend into the underlying river bed. Under such conditions the flow of the upstream water is impeded and a high hydrostatic pressure is developed. This pressure forces the water to break through to the surface, producing icings which may grown to a height of over 20 feet and may cause serious damage to the bridge.

The customary Russian method of bridge construction, consisting of cribbing framework sunk into the river bed and filled with stones has proved to be unsatisfactory. (Fig. 72). The cribbing filled with stones permits an easy access of cold air to a considerable depth below the river and causes deep freezing of ground, which, in turn, may start an icing even though no icing has heretofore appeared at that place. This difficulty can be remedied by alternating the layers of stones with some natural insulating material such as sod or peat. Cribbing piers with vertical sides are frequently heaved and tilted by river icings and icing-mounds. Pyramidal piers minimize or completely eliminate this deformation, as the lateral pressure of river ice tends to push the pier downward and thus makes it even more stable.

Wherever possible and practicable suspension bridges should be preferred.

Bridge pier consisting of cribbing filled with stones has been heaved and tilted 10°. Sosnovskoe, 161 km north of Skovorodino, Siberia. (From V.G. Petrov)

FIG. 72

Runways

Selection of a site suitable for a landing field is in all essentials subject to the same considerations as any other building site except that a level ground is sought and the site must be of larger dimensions. Because of the requirement for a large area the problem of site selection and runway construction is much more difficult than is the case of other projects. Large areas with satisfactory, homogeneous ground conditions are not very easy to find in the arctic and subarctic regions where the notoriously non-homogeneous glacial deposits form a good share of the terrain.

Runways may be installed in tundra country with preliminary drainage of excess of surface water and subsequent surfacing preferably with a non-rigid material (macadamized sand or fine gravel) as the surface is likely to suffer some warping which can be easily repaired from year to year.

Warping of the surface may be reduced by stripping off of the original silty soil and replacing it with sandy and gravelly layer prior to final surfacing.

Satisfactory runways may be built if the undisturbed surface ground is covered with a layer of low heat conductivity, such as pumice, slag, porous lava rock, air-entrained cement, etc. and then surfaced with cement.

Wooden runways have been built in tundra country with preliminary filling of the site with gravel. The gravelly base should be sufficiently thick to insulate the underlying permafrost and to prevent its thawing or recession. The thawing of permafrost will impair the stability of the base course and may cause local settling of the runway.

Wooden runways are constructed of 2" x 4" or 2" x 6" boards placed on edge and nailed together in a staggered order. The wooden pavement may rest directly upon the gravel or may be placed on criss-crossed 16 inch timber several feet apart with the spaces between the timbers filled with gravel and compacted by a roller. Pierced plank and other metal mats may also be used on a gravelly base.

Damages of surfaced road and runways are frequently caused by the heterogeneity of the subgrade material which cause a differential swelling and heaving. To minimize or to eliminate the effect of the differential heaving the subgrade material should be thoroughly mixed in order that it may be of uniform texture and composition.

Although runways on ice and water have no direct connection with permafrost their mention here is deemed justified in view of the general inaccessibility of the terrain and due to the need of emergency landings in the terrain where good landing fields are few and far between.

The aircraft equipped with skis during the winter and with floats during the summer cannot be operated in the early winter during the freezeup, and in the spring just preceding the breakup. During the freezeup it generally takes from 3 to 4 weeks for the ice to set in and to become sufficiently strong to withstand the load of a landing ship. The melting of ice in the spring and the clearing of the channel before the floats can be used may take 4 to 6 weeks. The length of the waiting period before the change from skis to floats and from floats to skis can be made depends upon the local geographic and seasonal meteorological conditions.

Generally speaking, the setting of ice will be slower in a body of flowing water, e.g., in a through lake; and much more rapid in a standing body of water. The reverse time relations hold true during the spring melting of the ice before the channel becomes open for the use of floats.

The spring melting of ice may be appreciably speeded up by coating the surface of ice with powdered charcoal, coal dust, cinders, or lamp black. This material may be scattered over the ice from an airplane. The dark particles having a high heat absorbing property will transmit the heat absorbed from the sun to the immediately adjacent mass of ice, producing gradually increasing cup-shaped depressions. As time goes on the surface of the ice assumes a rough honeycomb surface, with the dark particles lodged in the bottom of each depression. In high latitudes these depressions tend to assume an inclined position facing the low winter sun. A repeated application of black dust will melt the over-hanging ridges between the adjacent depressions and further speed up the melting. The use of charcoal is preferred to other powdered materials because of its light weight, which is a factor to be considered in transporting and spraying this material from an airplane. The charcoal will also tend to remain at the surface of what water may accumulate in each depression produced by the melting of the ice.

It is believed that the spraying of ice with oil will also bring about an accelerated melting of ice, but no experiments of this nature have yet been made.

In flowing bodies of water, e.g., in an inflowing and outflowing lake, the winter freezing of ice can be hastened by placing a boom across the current to check the flow of the surface water. The retarded surface water will freeze more quickly, and once the ice is formed it will continue to increase in thickness even though the water beneath it is in motion. The strength of the ice may be further increased by scattering straw and by pumping water over the surface of the ice. The small amount of straw embedded in the ice adds materially to its strength.

Experience shows that at the moment of landing an airplane loads the ice at the points of contact of the skis. This load may be divided into (1) static load (weight of the airplane) and (2) dynamic load (airplane impact at the moment of landing). Since airplane wheels or skis are designed to support a maximum six-fold overload at the moment of landing

(USSR safety factor), it is imperative that the ice be able to support a 600 per cent overload (Fig. 73).

When a landing must be made on young salt-water sea ice the above thicknesses must be increased 2 - 3 times depending on the area of the ice floe and its form.

However, the following facts must also be taken into consideration when operating ice airdromes:

(a) An ice layer is not perfectly elastic. A pressure for several hours (static load) produces substantial sagging. In the case of the railroad cars, this sagging at times reached the danger point.

(b) Dynamic loading did not produce the amount of sagging corresponding to that resulting from a protracted static load, because in dynamic loading the greater part of the force is spend in counteracting the resistance of water that is being expelled by the sinking of the ice, and in communicating a corresponding acceleration to this water.

(c) When an ice surface is loaded it is not just the loaded area that participates in the sagging, but also an ice field with a 25-30 m. radius around the area of loading.

(d) Experiments carried on by the USSR railroads have demonstrated that a 20 cm. thickness of fresh-water ice is required for partially loaded railroad cars (up to 15 tons) while a layer of freshwater ice 30 cm. thick is required for fully loaded cars (24 tons). When fresh-water ice had a thickness of 1.25 m. it was possible to move railroad cars by engine instead of by horses.

From the experience of arctic aviators the following conclusions may be derived:

1. No airplane should attempt to land on fresh-water ice with a thickness of 15 cm. or less nor, of course, on sea-ice of this thickness, which would be weaker.

2. When determining the possibility of landing on an ice-field it must be borne in mind that only a young ice cover which has been formed at temperatures below -9° C. is sufficiently strong to form a safe landing field. If young ice was formed at a higher temperature it will be necessary to wait until the temperature has stayed at -9° to -11° C. for several consecutive days; only then can young ice be considered strong and stable. This strength is not related to the thickness of the layer; thus, an ice layer as thick as 20 cm. formed at a temperature higher than -9° C. would not be safe for landing even a light airplane.

Fissures in the ice surface resulting from sharp temperature changes (from -2° to -27° C.) make flying operations difficult. Cracks on the runway wider than 25 cm. and ridges along the fissures higher than 10 - 15 cm. are dangerous for landing and take-off. Such fissures require

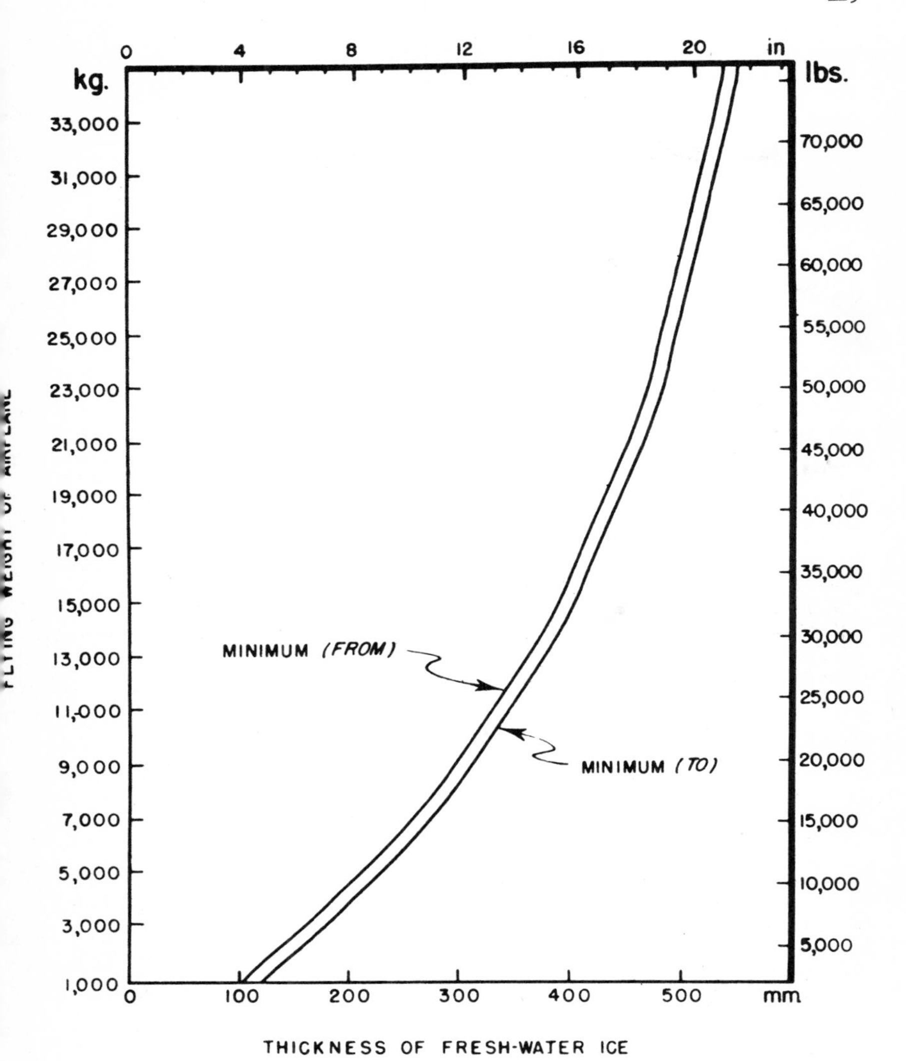

RELATIONSHIP BETWEEN FLYING WEIGHT OF AIRPLANE AND REQUIRED THICKNESS OF FRESH-WATER ICE.

FIG. 73

special attention such as artificial filling with water pumped through holes drilled in the ice.

It is recommended that ice airdromes be flushed with water to obtain a smooth surface. This should be done by hand or by motor pump. Where this is impractical, rollers should be attached to caterpillar tractors and rolled over the landing field.

Dam and Reservoir Construction

Practical experience in dam and reservoir construction in the frozen ground region is as yet very limited. Water stored in a reservoir will thaw the underlying and immediately surrounding ground causing it to cave and slide into the reservoir. If the underlying ground is porous the melting of the interstitial ice may start a serious leak. Under no conditions should a reservoir be built over ground that contains layers or lenses of ground-ice. In planning a dam or reservoir, it is, of course, important to know whether the area to be flooded is underlain by continuous permafrost or has islands of taliks. If taliks are present, their nature, origin, and extent should be carefully studies. The hydrology and thermal regime of the critical area should be surveyed in great detail.

Other Engineering Projects

Permafrost phenomena enter into consideration in many other fields of engineering work but the available information on methods employed and results achieved is as yet very meagre. Some special cases, however, merit mention as they illustrate the breadth of scope in human activity in which the frozen condition of ground enters into planning of various projects.

Cold storage vaults, for example, can be successfully constructed in the permafrost area. Perishable produce may be kept in chambers at constant below-freezing temperature and any desired temperature may be obtained by excavating a chamber at the corresponding level in the permafrost.

Another interesting example of the use of a permafrost phenomenon was related by P. S. Smith[5]. A normal behavior of the permafrost table was utilized to prevent spontaneous combustion in a storage pile of coal. A pile of coal, just as any road fill, especially if built during the cold season, and with a sufficient amount of moisture, will become permanently frozen below its surface layer and will thus avert spontaneous combustion.

5/ Personal communication.

Water Supply

Introduction. In the permafrost region, where swamps, lakes, and rivers are a prominent feature of the landscape and where terrain seemingly abounds with water, the procurement of adequate, the year around, supply of water may be a very difficult problem. Many rivers and lakes which can be counted on for the summer supply of water freeze through to the bottom during the cold winter season. Springs that may furnish sufficient amount of water during the summer not uncommonly stop to flow after the first winter frosts. Shallow wells which generally supply a small amount of poor to fair potable water commonly go dry or freeze during the winter. The most dependable source of water from deep wells presents a mechanical difficulty of drilling through the entire thickness of permafrost. The thickness of permafrost is variable from place to place, ranging from several feet to more than one thousand feet. There are inhabited places in the north where the melting of ice and snow is the only means of obtaining potable water.

Even where an adequate source of water is established a satisfactory solution of water supply and distribution still remains contingent upon design and auxiliary equipment properly adapted to cope with the extreme winter cold and the frozen condition of the ground. It cannot be overemphasized that in the permafrost area a preliminary survey of the ground should be much more thorough than is normally required in the area free of permafrost.

Unless an adequate supply of water can be developed at or near the site, the project will have to be moved to a more favorable location or else abandoned.

From the above it is apparent that in the permafrost region the question of water supply merits the foremost attention in a planning of a project.

Sources of Water. In the permafrost region, prospecting for water supply consists of either a reconnaissance or a detailed survey of existing water resources.

In a reconnaissance survey information is gathered on all the nearby sources of water (rivers, lakes, springs, aquifers) and the possibility of finding water below the permafrost is determined from the general geologic and hydrogeologic survey without resorting to drill-tests. Such preliminary surveys are made in the summer and again in the winter. If possible the summer survey should begin before the break-up of ice in the rivers.

Upon the completion of a reconnaissance survey the sources of water which are most promising and nearest to the place of utilization are subjected to a detailed investigation.

The purpose of the detailed study is to obtain full information on all the aspects of the source of water in relation to the demand. Furthermore the detailed study should furnish the basic data necessary for planning the development and distribution of water. Detailed investigation of a source of water should be carried on for at least a year and should always include "the critical period" in order to determine the yield of water during the least favorable part of the year. The critical period occurs near the end of winter when all the surface waters and certain types of ground-waters are at their minimum due to the maximum winter freezing. In the north this period is usually in May and is progressively earlier southward to southern Siberia where the critical period is in March.

The quality of water should be tested at least twice a year, before the break-up of river ice and the melting of snow and in the autumn during the maximum thaw of the active layer. Samples of water should not be smaller than 2 liters.

Normally the study of water resources is carried on contemporaneously with the survey of road routes, cantonment sites and other engineering projects as the final decision on the choice of a site must take into account the availability of a suitable supply of water. The sources of water may be divided as follows:

I. Meteoric water and snow.
II. Ground-ice and surface ice.
III. Surface waters (rivers, lakes), and
IV. Ground-water.

I. Meteoric Water and Snow.

Meteoric water may be utilized:

1. Directly from the run-off.
2. By gathering the rain and snow water from roofs and from specially prepared surfaces.
3. By condensation of the moisture in the air in specially constructed condensers.
4. By building reservoirs to conserve the run-off.

Of these only the last method is of practical significance. To assure successful construction and utilization of a run-off reservoir it is necessary to take into account:

1. Amount of precipitation (annual, monthly).
2. Evaporation.
3. Seasonal fluctuation of surface and ground-water.
4. The size of the basin (catchment area).
5. The most suitable location and capacity of the reservoir.
6. Geological and hydrogeological conditions of the dam site and the area that will be flooded.

II. Ground-ice and Surface Ice.

River, lake, or glacier ice has long been used as a source of water supply for small settlements. In the winter the blocks of ice are

CLASSIFICATION OF GROUND-WATERS IN THE PERMAFROST PROVINCE

Class	Type	Hydro-static pressure	Phase	Character of movement	Tempera-ture	Rocks (source)	Quality	Remarks
I Water above the perma-frost (Supra-permafrost water)	1/ Seasonal (freezes in the winter)	At times	Alternat-ing solid and liquid	(Gravita-tional-Artesian)	Alternating positive (above 0°C) and nega-tive (be-low 0°C)	Alluvium and un-consoli-dated deposits	Commonly turbid, easily contami-nated.	Generally unfit for human use but may be used in emergency.
	2/ Freezes in part during the winter	At times	In part alternat-ing solid and liquid. In part constantly liquid.	(Gravita-tional-Artesian)	Negative	Same as above	Less likely to be turbid and less easily contami-nated than above.	May be used for water supply but is not reliable as to the yield and sanitary con-dition.
	3/ Does not freeze in the win-ter	No pres-sure	Always liquid	Always gravita-tional	Constant-ly low negative	Same as above	Same as above	Same as above.
II Water within the perma-frost (Intra-permafrost water)	1/ Always liquid	Almost always present, rarely absent	Always liquid	A-Arte-sian B-Gravi-tational	Either always positive or always negative	Alluvium and un-consoli-dated de-posits, rarely solid rocks	A-Clear B-Fre-quently turbid	Suitable for water supply but are of rare oc-currence. Type A is preferred.
	2/ Always solid (ground-ice)	None	Always solid		Always negative	Same as above	Same as above	Cannot be used directly.
III Water below the perma-frost (Sub-permafrost water)	1/ Shallow-in allu-vial lay-ers, in fissures, and in solution channels (Karst)	As a rule constant, rarely absent	Always liquid	Artesian or sta-tionary. At times sealed by the perma-frost.	Low posi-tive or negative (if strong-ly mineral-ized)	In bed-rock, rarely in un-consoli-dated deposits	As a rule clear	Suitable for water supply un-less too strong-ly mineralized.
	2/ Deep-in layers (aquifers) in fissures, in solu-tion chan-nels (Karst)	Always present and con-stant	Always liquid	Artesian or sta-tionary	Always positive, occasion-ally quite high	In bed-rock.	Always clear	Suitable for water supply un-less too strong-ly mineralized. May be warm or hot.

stacked near the house and are used as needed. Locally ice is stored in cellars where it may last through the summer. On some polar islands ice is the only source of water as the running surface water freezes through and the ground-water along the coast is too strongly mineralized. In such areas ground-ice may also be utilized as a source of water.

III. Surface Waters (Rivers and Lakes)

In the permafrost area, rivers which can be used as source of water supply are not very numerous. During the winter most rivers freeze to the bottom except where the channel is deep, but even in these deep places water may become stagnant and unfit for use. Only the larger rivers and lakes furnish water throughout the year.

IV. Ground-water.

In the permafrost province ground-water is widely used as a source of water supply, especially in the southern part. In the north the thick layer of permafrost is a formidable obstacle to the development of the ground-water resources.

A. Ground-water from springs.

Springs may be fed dominantly by the water above the permafrost or dominantly by the water from below the permafrost. The first are of little practical value as they usually cease in the winter and their water is, as a rule, of poor quality. Some of these springs, however, derive their water from an aquifer (talik) which does not freeze in the winter. Such springs flow all through the year but they are relatively rare. On the other hand, springs which are fed by the water from below the permafrost form an important source of water supply.

Search for suitable springs should be carried out during the "critical period" at the close of the winter when surface ground is frozen to the maximum depth. Common surface manifestations of such springs are: seepages, icings, and other types of frost-mounds which form from underground seepages.

One or more springs chosen as the most suitable for the development of water supply are subjected to a detailed investigation during the following summer.

B. Water supply from above the permafrost (suprapermafrost water).

The water above the permafrost can furnish a constant supply of water only when the ground above the permafrost does not freeze to the bottom during the winter. Such conditions may be found along the banks of large rivers, in oxbow lakes, lakes with constant inflow and outflow, floodplains of large rivers and their lower terraces. The most common occurrence of unfrozen water above the permafrost is in the river bed below the bottom of the stream or near the edge of the stream. (Fig. 74)

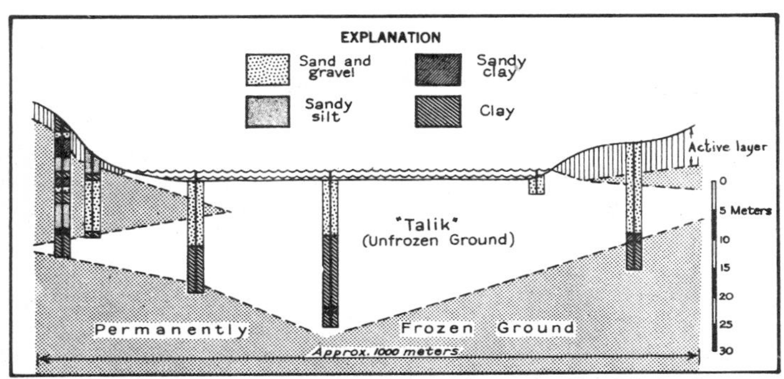

Cross-section through the Gorodskaya protoka (channel) at Yakutsk showing extent of frozen ground

(After Svetozarov)

FIG. 74

Unfrozen water above the permafrost may also occur at the mouths of valleys and at the heads of alluvial fans.

The study of maximum yield of this water is carried out in the autumn (September-October) when the active layer is thawed to the maximum depth.

Determination of the minimum yield of the water above the permafrost should be made during the "critical period" when active layer is frozen to the maximum depth. If the entire thickness of the ground above the permafrost is frozen, further investigation at that locality may be abandoned. It is always necessary, however, to make tests in the river bed, along the river banks, and in lakes. The yield of water during the winter should be determined by pumping for not less than 10 to 15 days, and should be accompanied by an analysis of the water. Prospecting for water above the permafrost should be supplemented with the systematic study of the regime of the aquifer fluctuations of the water-level, quality of water, temperature, rate of freezing and so on, and this information should be correlated with the meteorological data. An estimate should be made of the minimum yield during the most unfavorable year (subnormal precipitation, excessive frost, etc.). In prospecting for a dependable water supply, well organized test pumping should be continued from 1 to 2 months.

Shallow waters above the permafrost are easily susceptible to contamination and therefore require careful investigation and sanitary protection of the surface area.

Search for water which freezes during the winter may be carried out, in connection with some temporary project which is likely to last only through the summer, as for example, construction of a road or a temporary camp.

The most serviceable method of developing water above the permafrost is by a well or a trench.

C. Water supply from within the permafrost (intrapermafrost water).

Water within the permafrost most frequently occurs in thick alluvial deposits near rivers or near old river channels. It may also be found in rocks which are strongly jointed (fissured), along structural breaks (faults) or in rocks which are cut by veins and dikes. Water within the permafrost also appears to be present in alluvial fans and in areas with widespread ground-ice. It is frequently encountered near large springs which are fed by the water from beneath the permafrost. (Fig. 12).

The most widespread occurrence of the water within the permafrost is in the southern zone of the permafrost province. Here the ground is dominantly unfrozen and the behavior of the water is very similar to that of the ground-water outside of the permafrost province.

As yet no special methods have been developed in prospecting and studying of this type of water.

Water within the permafrost percolates through permeable layers or may flow freely through a pipe-like fissure. When this type of water is tapped care should be taken not to retard or interrupt the flow of water as slowing down of the underground circulation may result in freezing of the water-bearing layer. Excessive or accelerated pumping may also cause freezing of the underground water channel.

D. Water supply from below the permafrost (subpermafrost water).

Waters from below the permafrost are the most dependable sources of water supply. (Fig. 12). Their yield is usually large and, as a rule, of good quality.

Prospecting for water from beneath the permafrost should be preceded by a thorough hydrogeological survey of the area. This preliminary survey should determine the extent of the permafrost, its continuity, and its temperatures. Particular attention should be paid to large nonfreezing springs, wet icings, "hydrolaccoliths", existing deep drill-holes and to their relation to the geologic structure of the area. Water-bearing horizons which feed large springs and icings should be explored first.

Waters that occur at some depth below the permafrost are preferred to those which are found immediately below the frozen ground. These deeper waters are usually warmer and therefore, can be piped a considerable distance with less risk of freezing while in the pipes.

There are four types of water below the permafrost:

1. Alluvial
2. Layered (in bedrock aquifers)
3. Fissured (in joints)
4. Solution channels (Karst)

E. Alluvial water below the permafrost.

Alluvial water below the permafrost may be looked for in broad valleys of large rivers where geologic observations indicate that the alluvial deposits may be very thick and not completely frozen. There is little, if any, chance of finding alluvial water below the permafrost in valleys where these deposits are thin and where the thickness of the permafrost is great.

In the east-west trending valleys of the southern part of Eastern Siberia and in the Far Eastern region, the exploration work should proceed from the foot of the south-facing slope towards the middle of the valley. Test drills along the north-facing slopes usually give negative results.

In valleys of meridianal trend test drills should be placed near the river channel and extended to the side of the valley which receives greater solar radiation. If a seepage of water from beneath the permafrost is present at the foot of the slope prospecting for water should be started from that spring.

The minimum depth of drill-holes should be not less than the distance to the base of the permafrost. Most satisfactory results are obtained when the entire thickness of alluvium is pierced and the underlying bedrock is cut to a depth of one to two meters depending upon the character of the bedrock. The spacing of drill-holes is controlled by a number of local factors and generally vary from 50 to 200 meters. When the character of the alluvium and its water-bearing capacity is determined from this transverse line of drill-holes, the areal extent of the water-bearing alluvium is ascertained by making additional test drills up and down the valley at distances from 500 to 1,000 meters from the first line of holes.

Further investigation of the aquifer, its yield, quality of water, etc. is carried out in accordance with the general instructions given below. The points tentatively selected for exploitation should be tested by a prolonged test-pumping. Test-pumping should be done in February and March and should be accompanied by careful observations on the fluctuation of the water-level in the well, the regime of springs, etc.

The possibility of finding water below the permafrost in old-alluvial deposits on terraces should not be overlooked.

F. Water below the permafrost in bedrock aquifers.

In the permafrost region stratified bedrock may be drilled for water at a place where from geologic considerations there is a likelihood that water-bearing layer (aquifer) may be tapped below the permafrost.

For a synclinal structure test drills are placed along the axis of the syncline as well as across it, depending upon the lateral changes in the composition of the aquifer (change of facies). In general test wells are drilled in the center of a structural depression.

If several aquifers are present, the deepest one with "warm" water should be preferred. Warm water will be less likely to freeze during its passage through the overlying frozen ground when pumped to the surface.

Where the strata dip uniformly in one direction the well should be so placed that it will penetrate the aquifer at a level below the permafrost; preferably at some distance below its base as the water immediately below the permafrost is likely to be very cold and may easily freeze in the pipes while passing through the permafrost. Evaluation of water resources in bedrock aquifers should be made with caution and requires a thorough familiarity with the geology of the surrounding area. In areas where permafrost is continuous (Fig. 2) it is safer to count on a limited supply of water and not to expect a replenishment unless there is a good reason to believe that the aquifer is fed by some outside source. In such occurrences, in estimating the resources of water it is, therefore, necessary to compute the volume of the pore-space in the aquifer and the rate of underground percolation which will determine the rate of yield.

Where there are taliks or islands of permafrost it is important to determine what relation they have to the aquifer and whether or not the aquifer receives any meteoric water through the taliks. Under these circumstances a prolonged pumping test during the "critical period" is imperative.

G. Water below the permafrost in fissures.

In areas of fractured and fissured rocks open seams filled with water may represent:

a. Fissures or joints in the zone of weathering due to physical and chemical action of water, heat, or frost.
b. Fissures produced by tectonic action. (Fig.12).

The former are usually found at shallow depth and may contain an ample supply of water to a depth of from 30 to 100 meters. The latter are, as a rule, much deeper and are less common.

In the zone of joints and fissures, the ground-water usually has a continuous water table, whereas in the deeper tectonic fissures the water-bearing seams are disconnected and behave independently.

The formation of open fissures due to the process of weathering is most intense along the zones of tectonic breaks and along intrusive and other formational contacts. In areas of homogeneous rocks and undisturbed structures the processes of weathering are controlled by relief, location with reference to solar heat and many other factors.

In the absence of satisfactory methods of prospecting for ground-water in jointed or fissured rocks it is recommended that the exploration begin with natural seepages and springs which usually indicate the presence of water-bearing zones. The dependability of such springs should be verified by making observations through the year on their regime, their icings and, particularly, their yield during "the critical period".

If for some reason it is not feasible to develop a spring at the place of its original issue the water-bearing zone may be followed into the surrounding area and a more suitable site is selected for tapping this zone.

If no springs are found in the area prospecting for ground-water may be carried on by means of test drills or pits. The choice of places for these test holes should be guided by lithologic, structural, tectonic, and geomorphic considerations.

Favorable places for accumulation of water are along large fault zones, intrusive contacts, and along stratigraphic contacts. Very promising are contacts between coarse clastic sediments and the underlying eruptive rocks particularly if the latter have been subjected to deep weathering prior to the deposition of the overlying sediments.

Where outcrops are poor or altogether absent greater emphasis is placed on geomorphic indications of water.

It should be remembered that where the permafrost is only 30 to 50 meters thick practically the entire thickness of weathered and jointed rocks will be frozen through and will not yield any water. It therefore follows that, as in the greater part of northern Siberia, where the permafrost is considerably more than 50 meters thick, there is very little chance of finding any water in the fissured (jointed) zone of the bedrock.

More favorable conditions are found in southern Siberia, as for example, in Transbaikalia where permafrost is not continuous, being, for the most part, confined to the valley bottoms and commonly absent on the divides. In such an area the search for fissure water on the basis of geomorphic criteria usually gives satisfactory results. Drill holes placed at the foot of a slope in fissured (jointed) slates and other well consolidated rocks almost always produce water. Best results are obtained in the wells drilled at the foot of south-facing slopes. However one should not expect any large quantity of water from this type of occurrence in areas where the yield from individual wells is rarely more than 1 to 4 cubic meters per hour.

H. Water below the permafrost in solution channels (Karst water).

Abundant supply of water below the permafrost may be obtained from solution channels in extensive limestone areas, particularly where permafrost is not continuous. This water is especially widespread in the southern part of the Central Siberian plateau and in other limestone and dolomite areas.

Prospecting for water in limestone areas should be guided to a considerable extent by the geomorphic features of the region. It is important to know whether or not the surface openings in limestone (sink-holes, funnels, etc.) take in any amount of meteoric water and feed it to the channels below the permafrost. Where there is no such connection, no appreciable replenishment of the water supply can be expected and the total resource should be estimated solely on the basis of porosity of the rock, its structure and areal extent below the permafrost.

Test drilling in limestone should not be abandoned if no water is encountered immediately below the permafrost. In limestone the water-bearing channels may be very irregular and may occur at any depth below the base of the permafrost. As a last measure before abandoning a well a charge of dynamite should be set off at the bottom of the well. The explosion may shatter the surrounding rock and open the nearby water channels.

Construction of Wells. Shallow wells in unconsolidated sediments should be properly lined to prevent silting and caving. Well openings should be well insulated from the cold winter air, and preferably placed in heated buildings.

Ordinary wooden well casings are frequently deformed during the winter swelling of ground, especially in wells in low swampy ground and where the fill around the casing consists of clayey material. Wells should be placed on a high ground and the fill around the casing should be gravelly or sandy. The clay apron around the well, to prevent the surface water (spilled water) from leaking into the well, should begin a meter or a meter and a half from the casing. Presence of clay in contact with the casing is likely to cause heaving of the well casing.

In areas where the swelling of ground is very pronounced, the wells which tap water from above the permafrost should have a casing which cannot be easily pulled apart and out of the ground. The casing should be anchored in the permafrost to a depth of from 1 to 2 meters.

In operating a well, care should be taken to maintain the normal level of water as nearly as possible. A marked departure from this norm may cause a change in the hydrology of the ground-water and may result in the freezing of a well, or in a sharp decline of the yield of water. In coastal areas and on small islands excessive pumping is likely to cause infiltration of sea water.

Drilled wells have many advantages over dug wells. Their upper part is usually well protected from silting and pollution by the surface waters and they are, as a rule, less susceptible to deformation by swelling ground. Their main disadvantage is that the water in drilled wells easily freezes. The possibility of freezing is, however, considerably reduced if a large diameter pipe is used, if a deeper and warmer water is tapped, and if the pumping of water is not interrupted. Strangely enough excessive pumping of water, particularly of the water from within the permafrost, may also bring about the freezing of a well.

To prevent the freezing of a well while in the process of drilling a heated metal cable or an electro-thermal unit may be suspended in the well, and the drilling mud can also be warmed with steam. The use of salt water in the drilling mixture (mud) will likewise keep the well from freezing.

During a temporary suspension of the drilling operations the tools should come out of the well to avert the possibility of their freezing to the walls of the well.

Winter test-pumping should be done in an enclosed and heated pump house or tent so that the water in pipes and pump will not freeze if pumping is temporarily stopped. If pumping is stopped for an considerable length of time the pump and pipes should be drained.

If possible, wells should be sunk where permafrost is not very thick and where the temperature of the frozen ground is not very low. Around the southern periphery of the permafrost region the most dependable wells are those which are drilled through a permanent talik.

Laying of Water Pipes. In the permafrost region one of the most difficult water supply problems is the laying and upkeep of pipes. Formerly water pipes were laid in conduits or tunnels which were heated by either steam or hot water, but conduits themselves are frequently damaged and are difficult to repair. Modern technique, based chiefly on the work of Chernyshev, Tsytovich and Sumgin, offers satisfactory means to overcome these difficulties.

Water pipes are commonly buried 2 or 3 meters below the surface of the ground and are imbedded either in the lower part of the active layer or in the upper layer of the permafrost. Pipes should be buried deep enough so that the winter extremes of temperature will not affect them. A talik between the active layer and the permafrost would provide an ideal ground for pipe-lines.

In laying water mains, ground with a prolonged zero-curtain condition should be preferred. (See page 17). A zero-curtain condition can be created around the pipe by imbedding it in a moist ground (clay) with low permeability. This moisture-bearing packing (clay) should be surrounded with a low heat-conducting material such as peat, moss, or some commercial product. (Fig. 75).

According to Chernishev the heat balance of the ground containing pipes should be regulated in such a way that during the summer the

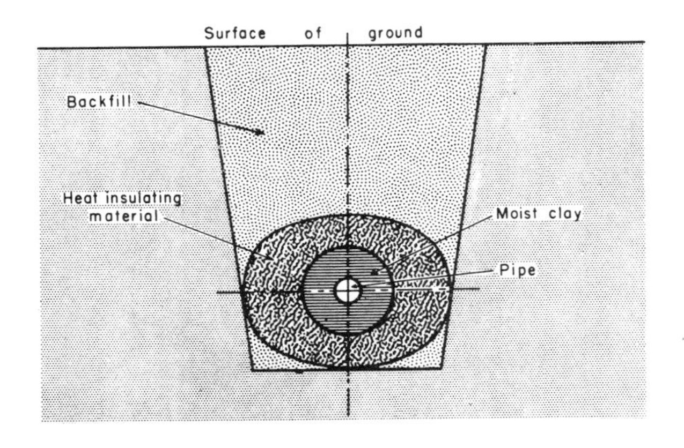

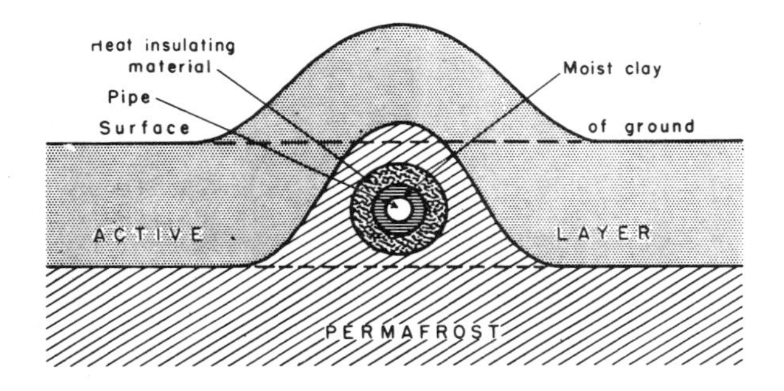

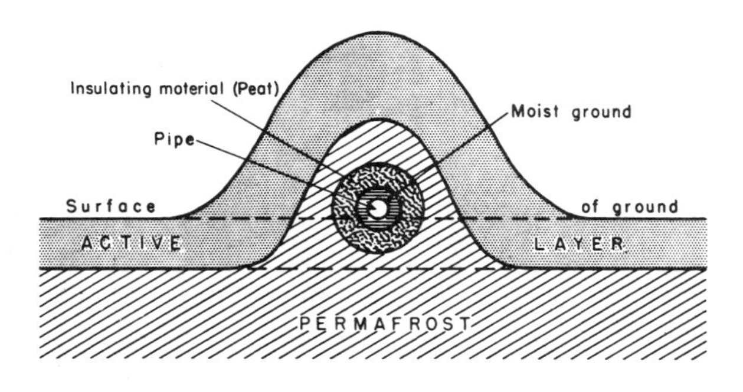

Laying of water pipes in the permafrost area

(From Tsytovich and Sumgin)

FIG. 75

covering ground should readily conduct the heat downward; during the winter, when the heat goes in the opposite direction and is lost through radiation, heat conductivity should be reduced. The most satisfactory ground for this purpose is sand. By wetting the sand in the summer and by drying it before the beginning of frosts it is possible to maintain the desired heat balance of the ground. According to some reports this method is not entirely satisfactory and is more costly than the others.

Pre-heating of water at the pump-house, at the head of the distribution system is a satisfactory way to keep pipes from freezing. Buildings and establishments which are likely to give off waste steam or other heat should be located near the source of water so that the available heat can be utilized for warming the entire system.

Pipe-lines should be so planned that the buildings which are expected to consume the largest amount of water are located at the end of the water main - farthest from the source. In this way a maximum flow of water will be maintained in the pipes, reducing the possibility of freezing.

Burying of water mains near sewer pipes will substantially simplify the problem of anti-freeze measures.

The Excavation of Ditches During the Winter. The procedure followed in the laying of pipes during the winter is entirely different from that which is employed during the other seasons. The greatest difficulty is encountered when the exposed pipes, during the cold winter months, are being tested for pressure. The excavation of ditches, the laying of pipes, and the connection of joints during the winter months does not present any particular difficulties. As a matter of fact, under certain climatic conditions it is even more advantageous to carry out this work during the winter than during the warm summer months when an excessive precipitation may materially interfere with the progress of work. The ditches dug during the summer frequently require cribbing or other reinforcement of sides to prevent them from slumping and necessitate pumping of water by manual or mechanical means. If, however, the digging of ditches is carried out from early winter to early spring, there is no need of reinforcement and there is no problem of drainage. The excavation work during the winter will naturally require additional personnel and involves some expense caused by the difficulty in breaking the frozen surface ground. However, this additional expense is for the most part, more than offset by the economy which is achieved by the saving of materials and personnel that would otherwise be required on the job during the summer.

Other considerations which enter into the planning of the water pipelines are the possible delays in the operation of hydro-electric plants and in the function of other installations which are dependent upon the projected water supply. Another advantage of winter construction also lies in the fact that a greater supply of manpower is generally available during that season.

The excavation of ditches for water mains should be preceded by a survey to determine the thickness of the frozen ground. The thickness of frozen ground may vary, not only from place to place, but also from time to time during the construction.

Explosives may be used to an advantage in the excavation of trenches. The holes drilled for the explosive charges are generally placed in an inclined position to produce a lifting action with the blast. The underlying thawed ground, if present, is excavated in the usual manner. In a given region with the advent of the frost, there is usually no interference from the rain. The sides of the ditch will require no bulkhead and even the thawed ground beneath exposed to the cold air, will freeze and hold in place.

Heat Computation of the Water Distribution System. Water pipes in permafrost or in ground that is subject to seasonal freezing are protected from the frost by the warming of water directly at the pump or at the distribution point. Where lines are very long intermediate points of warming of water should be established. The amount of heat added to the water is so calculated that the water upon passing through the entire system and after remaining static during the maximum allowable time should not drop below a certain permissible minimum temperature.

After the water pipes are laid and backfilled the temperatures in the surrounding ground become stabilized in such a manner that the isotherms (points of equal temperature) roughly parallel the surface of the ground. (left half of the diagram, figure 76).

After the water system is put into operation the ground around the pipe receives a certain amount of heat from the flowing water and the ground isotherms assume a different position, which is shown in the right half of figure 76. With the continued operation of a water system the temperature of the ground adjacent to the pipe will be more or less constant. Minor changes in operation may be caused by the variation in the pre-heating of water, the duration of uninterrupted operation, the velocity of the flowing water, as well as by the general climatic changes. In computations the temperature of the ground which surrounds the pipe is determined by direct measurements prior to the installation of the pipe, or is obtained from that data furnished by local meteorological stations. In view of the fact that during the year the ground at a depth of 2½ to 3 meters changes within certain limits, the minimum values are generally used in computations.

The loss of heat in a water system depends to a considerable degree upon the operation of the system. The most intensive loss of heat is observed during the flow of the water when the transfer of heat takes place through CONVECTION. A very small amount of heat is lost during the static period when water does not move through the pipes, at which time the transfer of heat takes place through CONDUCTION through the peripheral layers of water.

Quantitive values of coefficients of the loss of heat K_1 and K_2 obtained empirically for pipes in clay ground are given in the following table:

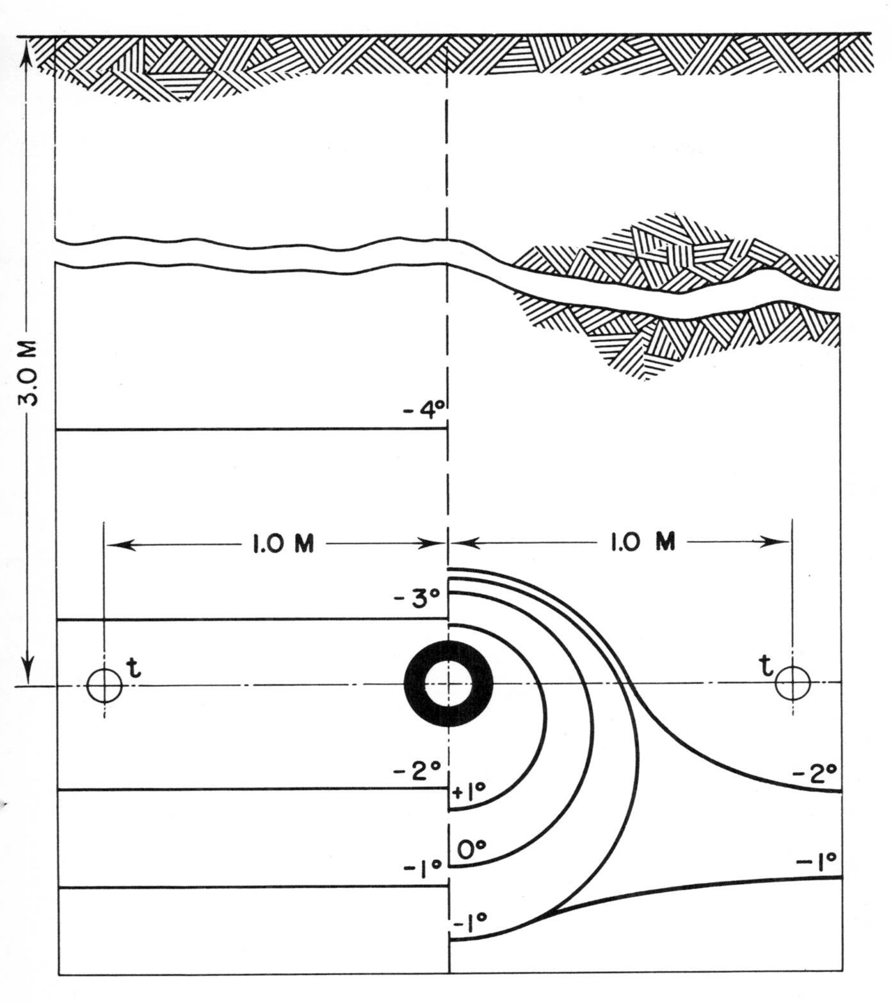

Before the flow of water During the flow of water

Ground isotherms near pipe

(DEGREES IN CENTIGRADE)

FIG. 76

The Values of Coefficients K_1 and K_2 cal/m^2 per hr. per °C. During the Normal Operation of the Distribution System

Operation of the System	Kind of Pipe	K_1 - During Flow of Water	K-at the start of operation	K_2 During Non-flow period	K-at start of non-flow period
With the daily interruption in flow of water	Cast Iron	10	20	1-2	4
	Wooden Stave	3.6	6	1-2	3.5
With continual flow of water	Cast Iron	3	20	-	-

The nomogram in figure 77 enables one to determine with sufficient accuracy the temperature of water in a distribution system. This nomogram is constructed on two adjacent and interrelated grids.

On the left half of the graph:

L - Represents length of the water system in meters.
d - Diameter of pipes in millimeters.
v - Velocity of the flow of water m/sec.
Z - Duration of the static period of water in pipes in hours.

On the right half of the graph:

K - Coefficient of the loss of heat.
τ_1 - Difference in temperature between the flowing water and the ground at the starting point of the system.
τ_2 - Difference in temperature between the flowing water and the ground at the terminus of the system.
τ_2' - Difference in temperature between the standing water and ground at the starting point of the system.
τ_2'' - Difference in temperature between the standing water and the ground at the terminus of the system.

The key explaining the use of the nomogram is shown in the lower right-hand corner of the nomogram.

Example. Given: L = 5000 m; d = 150 mm; v = .60 m/sec; Temperature of water at the starting point of the system, $T_1 = 10^\circ$ C.; Average temperature of ground during the static period, t, equals -3° C.

Determine temperature of water at the terminus of the system, T_2, and the maximum duration of the static period Z for the temperature of water at the end of the static period $T'' = 2^\circ$ C.

Solution:

$$\tau_1 = 10^\circ - (-3^\circ) = 13^\circ$$

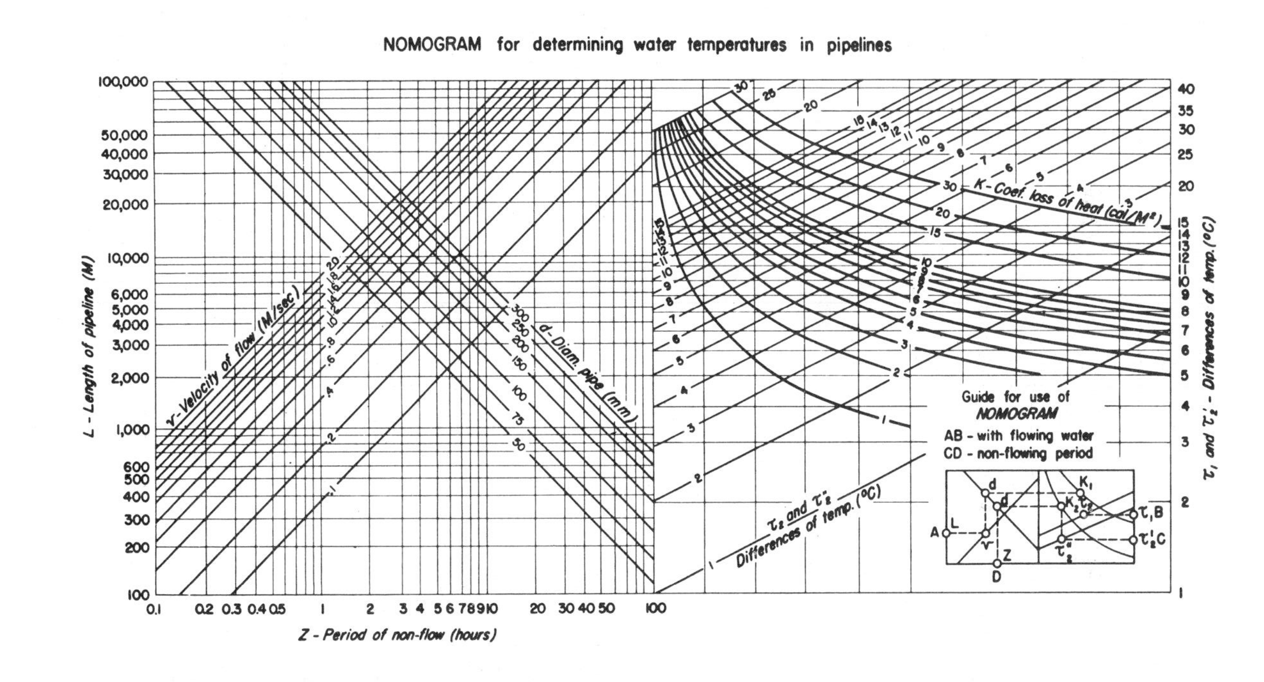
NOMOGRAM for determining water temperatures in pipelines
L - Length of pipeline (M)
Z - Period of non-flow (hours)
v - Velocity of flow (M/sec)
d - Diam. pipe (mm)
K - Coef. loss of heat (cal/M²)
τ₁ and τ₂' - Differences of temp. (°C)
τ₂ and τ₂" Differences of temp. (°C)
Guide for use of
NOMOGRAM
AB - with flowing water
CD - non-flowing period

FIG. 77

Using the nomogram for $K_1 = 10$, we get $\tau_2 = 7^{\circ}$; therefore

$$T_2 = \tau_2 + t = 7^{\circ} + (-3^{\circ}) = 4^{\circ}$$

$$\text{with } \tau_2 = \tau_{\frac{1}{2}} = 7^{\circ}$$

$$\tau_2'' = T'' - t = 2^{\circ} - (-3^{\circ}) = 5^{\circ}$$

with $K_2 = 1$, we obtain from the nomogram Z = 12 hours.

The values of the coefficient K_1 and K_2 given in Table, page 146 correspond to the normal operational conditions of a system, which is kept in operation every day with a regular daily flow of water. Under these conditions the consumption of water should not be less than 8 times the volume of pipes of the entire system.

In addition to the computation of heat values for a normal operation of a water system it is also necessary to calculate the heat exchange during the initial stage of a system when water is turned on for the first time.

When water is turned on at the completion of the installation and after a prolonged period of inactivity or after the pipes had been drained the ground surrounding the pipes attains the lowest temperature and the pipes are appreciably chilled. It therefore follows that when water is turned on again there is a considerable absorption of heat by the pipes and by the immediately adjacent frozen ground resulting in a substantially higher coefficient of the loss of heat (K) than during the normal operation. In every water system, therefore, one should take into account the need of additional heating of water during the period when the system is first put into operation, The coefficient K is as yet based on only a few experiments and is given in the table on page 146.

When water is turned on in a newly constructed water system, it is sufficiently safe to allow the temperature at the terminus of the system (T_2) to drop to between 1° and 2° C.

This temperature at the terminal point will prevail only during the initial stage of operation of the system. With the passage of water throu the system, this temperature will gradually rise and after a time will attain a certain stability. After that the water system may be allowed to remain static for a duration in accordance with the computed values for such a condition.

The temperature of a distribution system generally becomes more or less stabilized after the volume of water that had passed through the syst is about twenty times the volume of the entire pipe system.

In order to speed up the warming of the pipes during the initial stage of operation the water is generally forced to flow through the pipes at a higher velocity, or, if facilities are available, the water at the starting point is warmed up more than is required for the normal operation. It should be borne in mind, however, that should the water be warmed up to a temperature above 20° C., some damage may be caused to the pipeline by the excessive expansion, particularly at the joints.

Under normal operation of a distribution system the temperature of water at the terminal point (T") during the static (non-flow) period is generally maintained between 3° and 5° C.

Two fundamental principles should be used as the basis for the calculation of the heat values in a water distribution system:

1. For determining the temperature at the terminal point of the system, the temperature in the component sections should be calculated in a consecutive order in the direction from the terminal point towards the intake where the water is being warmed up.

2. Knowing the available amount of heat from the heating unit the temperature of water (T_1) is determined after the water leaves the heating unit. Then the temperature of water is verified in a consecutive order at all the distribution centers down to the terminus of the system. The terminal temperature should never be allowed to drop below the computed permissible minimum.

<u>Pressure Test of Water Distribution System</u>. Before any water distribution system is put into operation all pipelines of the system are subjected to a pressure test. Special conditions arising from the freezing temperatures of the air require a particular procedure in the carrying out of the pressure tests. Pressure tests of each water system consist of:

1. Filling of pipes with water.

2. Subjecting the water to pressure.

3. Draining the pipes.

It is apparent that during the entire process the temperature of water should not be allowed to drop below the freezing point. It should also be remembered that water has the highest specific heat of all the substances and that it passes from the liquid state into solid - ice - after it is cooled to a temperature below zero degree Centigrade, and after all the latent heat of fusion is expended. The loss of heat by water depends in large measure on the condition of water. While the water is moving the loss of heat takes place dominately through convection, whereas during the static condition of water, the loss of heat is reduced and takes place only through the peripheral layers of water.

The water having been warmed up prior to entering the pipes may be kept at a temperature above the freezing point throughout the test, thus preventing its freezing in the system. The greatest chilling of the water will take place at the moment of its initial filling of the pipes. A considerable amount of heat will be absorbed from the flowing water by the cold surfaces of the pipe, and will pass into the adjacent ground. The first portion of the water, by the time it reached the end of the pipeline, will have a temperature close to zero degree Centigrade. For this reason it is dangerous to allow this water to remain in the pipes for any length of time. The water should flow through the entire system of pipes and drain out at the terminal points, the temperature of which will gradually rise above the freezing point.

Preliminary warming is generally required before the water is allowed to flow into the pipes. The time interval during which the water, while passing through the pipes, will chill down to freezing point should be carefully calculated. This is a prerequisite to a timely termination of the test to avoid the possibility of freezing of pipes.

Pressure testing of a water system should be carried out by sections of the entire length of the pipeline. The length of each section depends on:

1. Longitudinal profile of the line.
2. The efficiency of the warming up installation.
3. Means of delivery of water to fill the pipes.

Turning on of Water in a Water Distribution System. As has been stated above, the greatest loss of heat in a water system takes place during the initial stages of operation. Experience has shown that in a newly constructed water system the greatest number of failures, due to the freezing of pipes, happens immediately following the turning on of water. The freezing of pipes may be due to several causes:

1. Plugging of pipes by extraneous objects accidentally left in the pipes during their installation.
2. Faulty drain valves at the low points of the system failing to let the water out quickly in the case of local plugging of pipes.
3. Failure of pumps which may cause a prolonged stoppage of the flow of water.
4. Insufficient warming of water.
5. Stoppage of pipes by ice plugs in the sections which were not thoroughly drained after a hydraulic test.

In view of the above, a thorough inspection of the entire distribution system should be made before the water is turned on. In the systems which remain inactive for a long period of time, the ground adjacent to the pipes generally becomes chilled and the low places in the pipeline often become plugged with ice. These ice plugs may arise from the condensation of the moisture from the air in the pipe, or by the concentration of film water left in the pipe subsequent to the draining. The presence of plugs in a pipeline between two adjacent manholes may be detected by sound signals. A voice will carry well through the distance up to 1,000 meters. Should the voice fail to carry, or be inaudible, it is a sure sign that the pipe is plugged. The exact location of the plugged spot may be ascertained by a flexible, spirally wound steel cable called "snake". Should the "snake" pass freely through the section of pipe under inspection, the ground is excavated and the pipeline is exposed near the middle point. A small hole is drilled into the pipe and the sound signals are repeated. It is thus possible to determine which half of the pipeline section has the obstruction. The half that fails to carry the sound signals is in turn exposed in the middle, drilled through and checked as before. In such a manner, the inspector will ultimately reach the spot with the obstruction.

The checking of pipes may be also carried out by blowing of compressed air or steam.

It is often difficult to thaw out the pipes without going to the trouble of excavation and taking the pipes apart. The thawing of pipes may be done by means of bonfires arranged along the pipeline or by a jet of steam. Under no circumstances should the cast iron pipes be thawed out by insertion of heated metal rods or hot steam, as these will invariably damage the pipes. Wood staves may be thawed out by warmed-up metal rods or by a forced flow of hot water or steam. Bonfires should not be built in close proximity to wood staves as the heat from the fire is likely to "melt" or burn the asphalt coating of the staves.

In starting a water distribution system, it is necessary to warm up the ground adjacent to the pipes by a prolonged and intensified pumping of water through the system and by allowing a free flow of water through the terminal outlet valves. This warming up operation usually lasts from three to five days. As time goes on, the water may be left standing in the pipes for progressively longer periods of time. The duration of the warming up period until the temperatures are stabilized may be determined by the nomogram on page 147.

In starting an operation of a very long distribution system with very cold pipes and surrounding ground, the heating devices designed for normal operation do not furnish a sufficient amount of heat to permit the flow of water to the terminal points without the danger of freezing. Such a water system should, therefore, be put into operation by sections. After the first section is warmed up so that the temperatures have been stabilized, the next section is then turned on and so on. After a prolonged warm-up pumping of water, the system may gradually be brought to a normal operational schedule by shutting off the pumps progressively longer periods of time each day.

Water mains installed during the warm season of the year, backfilled with the thawed ground and put into operation prior to the deep penetration of winter frost do not require any preliminary measures of warming up and can be turned on in the usual manner.

To maintain a normal operation of any water system and to keep track of the temperature conditions, it is necessary to install thermometers at the beginning and at the terminal points of the system so that the temperature of water in pipes and the temperature of the adjacent ground may be observed. Fig. 78.

Water meters should be installed at all feeder mains and at other critical places along the distribution system in order to furnish the data necessary for the computation of the loss of heat and to furnish such other control data that may be required for a satisfactory or rational operation of the system.

The operational record of a new water system should contain the following information:

1. Duration of pumping.
2. Total flow of water.
3. Temperature of water in the well or at the source or intake.
4. Temperature of water at the initial point of the distribution system.
5. Temperature of water at the terminal points.
6. Temperature of ground near the pipes.

These records are invaluable to insure an adequate and the most rational operation of the system.

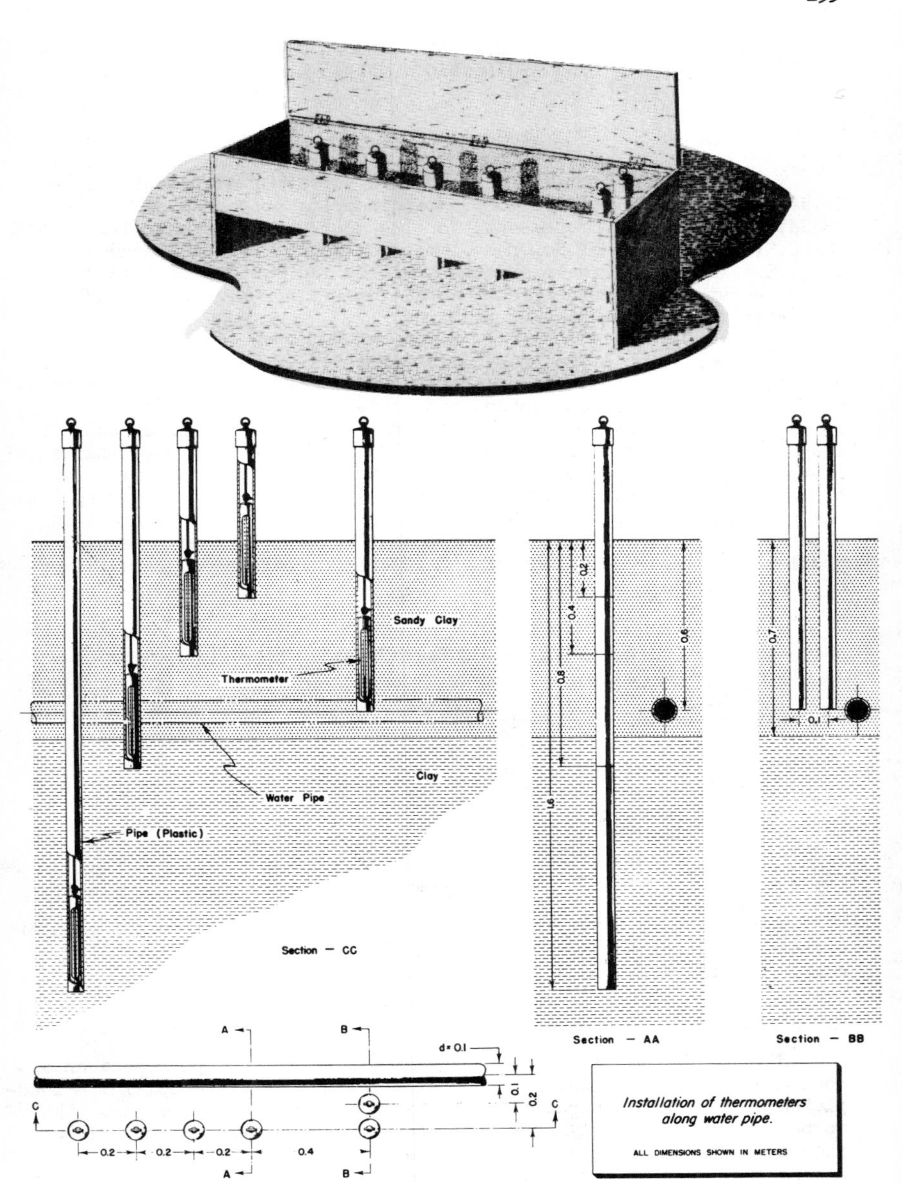
Sandy Clay
Thermometer
Clay
Water Pipe
Pipe (Plastic)
Section — CC
Section — AA
Section — BB
0.2
0.4
0.6
0.8
1.6
0.7
0.1
d = 0.1
A
B
C
0.2
0.4
Installation of thermometers along water pipe.
ALL DIMENSIONS SHOWN IN METERS

FIG. 78

INVESTIGATION OF FROZEN GROUND IN CONNECTION WITH ENGINEERING PROBLEMS

Introduction

The following outlines for study and survey of different permafrost phenomena, as adapted to such engineering projects as buildings, roads, bridges, etc., represent, for the most part, recommendations of the U.S.S.R. Acadmey of Science Committee on Frozen Ground headed by M. I. Sumgin. These recommendations embody the results of recent intensive study of frozen ground conditions in northern Russia and represent the current Russian practice.

The completion of comprehensive permafrost surveys may require as much as a year or more. Under the pressure of the existing war conditions this may be impracticable, but, notwithstanding this shortcoming, it is deemed necessary to present the recommended outlines in full for they are intended to serve as a check list, leaving it to the individual executive engineer to modify the procedure and to adopt the plan appropriate to each problem. It should be borne in mind, however, that the taking of shortcuts in the recommended procedure might result in leaving out of consideration factors that are critical to a particular problem with the possibility of subsequent partial or complete failure of the project.

General Survey of Permafrost

The study of the permafrost problems resolves itself into the following six main and closely inter-related elements:

1. Topography
2. Earth materials
3. Hydrology
4. Climate
5. Botany (ecology)
6. Laboratory tests and experiments

A preliminary survey for any project cannot be regarded as completed unless all six of the above mentioned elements have been carefully examined, analyzed, and correctly evaluated.

I. Topography:

With other conditions being equal, the permafrost phenomena on a mountain side will differ from those on the adjacent floodplain. Floodplains with general characteristics of impeded drainage are more likely to be underlain by silty ground and to contain ice lenses and wedges whereas in a well drained basin, gravelly ground is of more common occurrence. Conditions on a south facing slope will be materially different from those of a north slope, and so on. In some localities recognition of certain minor but characteristic topographic features enables a trained observer at a glance to evaluate the terrain as to the relative magnitude of permafrost problem in that particular area. Interpretation of topography should be assigned to a geologist, preferably one with some experience in the study of land forms (geomorphology).

II. Earth Materials:

Materials affected by the frost action include soil, unconsolidated clay, silt, sand, and gravel and other more or less consolidated bedrock. The study of earth materials should be carried out jointly or separately by a geologist and soil engineer. Permafrost mapping and soil survey, preferably with a geologic map as a base, should precede ground exploration work, and in turn, should be further augmented with the exploration data.

The work of a geologist and a soil engineer will include the study of composition, thickness and structure of the seasonally and permanently frozen grounds as well as the areal distribution of each specific kind of ground. Texture and permeability of ground should be observed in the field and supplemented by laboratory studies.

The following physical tests of the soil samples should be made in a laboratory by the soil engineer: grain analysis of the sample, moisture contents, specific gravity, liquid limit, plastic limit, plasticity index, consolidation test, shrinkage limit, natural density, California bearing ratio, plate bearing test, compaction test and others.

Ground temperatures should be taken at different levels and at different times of the year. An effort should be made to correlate the type and the distribution of vegetation with the change in the permafrost conditions and the character of the soil. It is very desirable to have a botanist-specialist in plant ecology (relation of plants to their environment) - to accompany a permafrost or soil survey party.

The results of the combined study of a geologist, soil engineer, and possibly botanist should be compiled and graphically presented in what may be called a "permafrost map". An example of a permafrost map is reproduced from Yanovsky. (Fig. 79). This map clearly differentiates two types of terrain - two terraces. The lower terrace appears to be very unfavorable. The surface of this terrace is covered with frost-mounds and there is a large icing at the foot of the second terrace. As can be seen from the type of vegetation, the ground of the active layer is oversaturated with moisture. The frost phenomena present on this first terrace are not found on the higher terrace. Judging from the vegetation, the ground of the higher terrace is drier and is well drained. These conclusions, according to Yanovsky, were later substantiated by the data obtained in test pits.

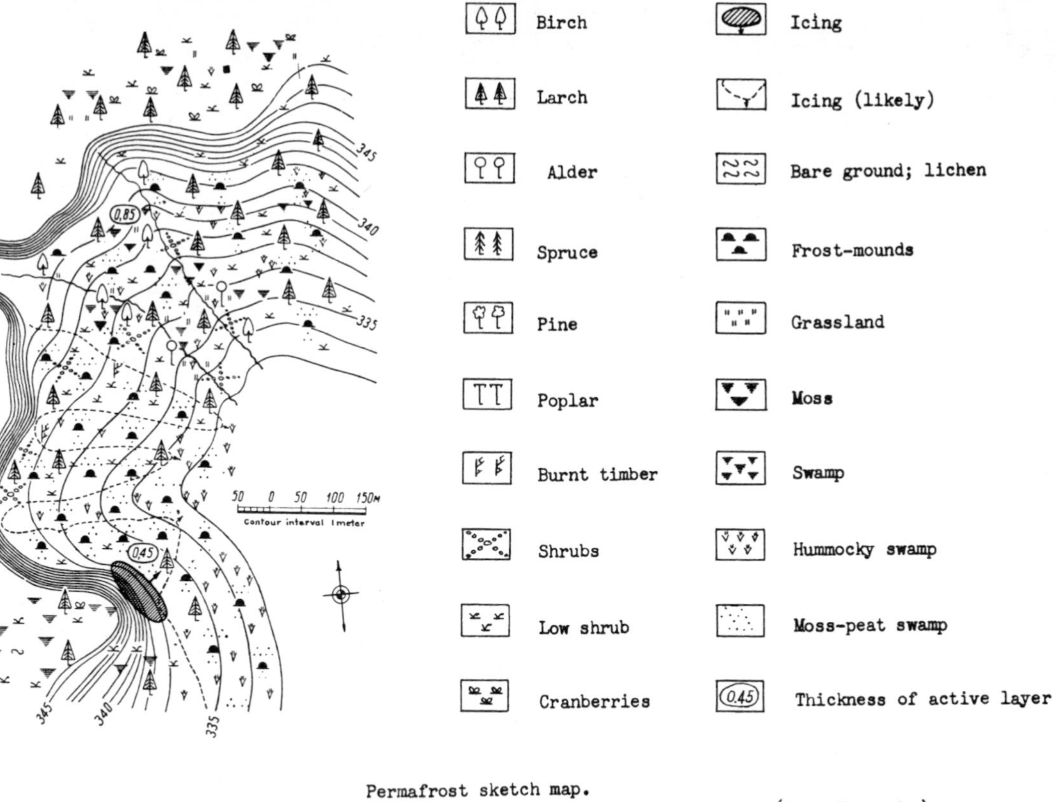

Permafrost sketch map.

(From Yanovsky)

FIG. 79

III. Hydrology

The investigation of water resources and of ground-water conditions should be entrusted to a hydrologist who is also capable in handling the surface water problems. The hydrologist should work hand-in-hand with the geologist. The two will study jointly the character and thickness of the seasonally frozen ground (active layer), constancy of this thickness over an area, moisture content and permeability of the ground, thickness of the permafrost, depth to the permafrost table, vertical profile of the permafrost, unfrozen layers or pipes ("taliks") at the surface, within the permafrost, and beneath the permafrost, ground ice, its mode of occurrence, thickness, etc.

IV. Climate

All the above mentioned men will draw upon all the available climatological data which should be furnished to them by meteorological and weather stations. The scope of the already existing meteorological stations should be expanded to include, in addition to the usual observations, the following:

a. Temperature of the ground. Daily recording of ground temperatures at various depths below the surface several times a day and in varied landscape surroundings (open areas, wooded area, sandy ground, clayey ground, under paving blocks, areas with heavy snow cover, areas with little or no snow, etc.). It is desirable that at least one or two places be selected where ground temperatures should be taken at intervals down to the depth at which there is no annual fluctuation of temperature.

b. Solar radiation. To measure the amount of solar radiation (actinometric energy) by means of sensitive instruments such as heliographs or actinometers. The effect of solar radiation on heat absorbing surfaces (runway pavements, buildings, accumulated effect of direct radiation plus the reflection from the sun-facing walls) should enter into computation of the thermal regime of the ground underneath or immediately adjacent to buildings and pavements.

c. Gaging of streams and other periodic observations on rivers, lakes, springs, thickness and duration of ice, freeze-up, breakup, etc.

It is also recommended that one or several buildings in a typical permafrost environment be chosen for a careful study of the frozen ground beneath them. Complete record to be kept of the design and the materials used in these structures and of the amount of heat generated in their various parts. Periodic observations of the ground temperatures are to be made at frequent intervals underneath the different parts of these buildings. Such experimental buildings will furnish much valuable information which will materially contribute towards a satisfactory design, construction and a successful maintenance of structures in the future.

To expedite the observation, measurements and the recording of the permafrost data at each station it is recommended that a technically qualified man, recorder, possibly in the Engineer's office or from the weather station personnel be placed in charge of this work. This recorder should be familiar with the main aspects of the permafrost phenomena and should have at his disposal at least two men skillful in handling simple soil testing tools.

The essential climatological data may be summarized in a tabular form as shown on page 160.

V. Botany

Geologist or soils engineer, preferably a botanist-ecologist should note the commonest plants, their associations and areal distribution. This information should be correlated with the permafrost condition of the terrain as revealed by the soil profiles in test pits and bore holes.

VI. Laboratory tests and experiments

A field laboratory should be established on the site of every major project so that soil samples, especially undisturbed samples, would suffer the minimum of damage or change of condition while in transit between the test pit and the laboratory.

All basic soil analyses and tests should be done at the site so that the results may be made available to the field men as soon as possible to guide them in their field investigation.

It is also desirable to have a well equipped laboratory at some central location where more elaborate tests and experiments could be carried out. For example, there is still much to be learned about the heat transfer in different soils; the heat conductivity of different building materials; the capillary water; the behavior of various water repellent lubricants, etc.

The data obtained by a geologist, soils engineer, hydrologist, and meteorologist are then re-worked, computed, and coordinated by a physicist-engineer and are submitted in a finished form (tables, graphs, equations) to a civil engineer who embodies this information in a design of a projected structure.

The following table gives a brief resume of the plan of study, the division of labor, and the general scope of the permafrost problem.

Scope and Plan of study of permafrost

Personnel

Geologist — TOPOGRAPHY - Mountain slopes, floodplains, river valley, alluvial fans, old lake basins, river and lake terraces, coastal plains, glaciated areas, frost-mounds, polygonal soils, slumps and landslides, geographic location, orientation with reference to compass points, etc.

Soil Scientist; Botanist, ecologist — MATERIALS - Field Work:

Mineral and texture composition
Thickness
Structure
Porosity (permeability)
Thermal regime of ground
Plant ecology
Thickness of active layer
Constancy of the thickness of active layer
Ground ice, its thickness, extent, etc.
Thickness of permafrost
Permafrost table
Vertical profile of permafrost
Thawed ground (taliks)
Mapping of permafrost phenomena
Soil Survey (mapping)
Soil Profile
Laboratory tests of Soil Samples:
(1) Grain analysis
(2) Moisture contents
(3) Specific gravity
(4) Liquid limit
(5) Plastic limit
(6) Plasticity index
(7) Shrinkage limits
(8) Natural density
(9) California bearing ratio
(10) Plate bearing test
(11) Compaction test etc.

Meteorologist — CLIMATOLOGY - Temperature of the air (and ground)
Precipitation (rain and snow)
Distribution of precipitation in time
Amount of solar radiation
Cloudiness
Humidity and evaporation
Winds, etc.

Hydrologist — HYDROLOGY - Ground-water: above the permafrost, within the permafrost, below the permafrost.
Surface waters: rivers, lakes, springs, etc.
Sea water (contamination by and thermal effect of)
Thermal regime of ground-water, surface water, and sea

Recorder (One at each station).

SUMMARY OF CLIMATOLOGICAL DATA

Geographic position: ____________________
(or name of meteorol. station)

Longitude: ____________ Latitude: ____________ Altitude: ____________

Name of nearest town: ____________ Distance to it: ____________

Name of drainage basin: ____________________

	Mean monthly												Mean Annual
	J	F	M	A	M	J	J	A	S	O	N	D	
Barometric pressure													
Winds: direction													
velocity													
Temp. of the air													
Minimum temp. of the air													
Maximum temp. of the air													
Absolute minimum													
Absolute maximum													
Days with freezing t°													
Precipitation													
Minimum precipit.													
Daily precipitation													
Mean minimum													
Mean maximum													
Days with precipit.													
Humidity of the air													
Mean minimum													
Mean maximum													
Evaporation													
Mean minimum													
Mean maximum													
Dew point													
Thickness of snow-cover													
Beginning of " "													
Disappearance " "													
Cloudiness													
Sunshine													
Solar radiation													
Soil temp.: at surface													
at ft.													
at ft.													
at ft.													
at ft.													
at ft.													
at ft.													

Planning of the Permafrost Survey

In the Arctic and sub-Arctic regions the usual engineering methods and practices cannot be successfully employed. Costly experience has shown that it is uneconomical if not futile to "fight" the natural forces of frost by using stronger materials, more rigid designs, or to resort to periodic and excessively costly repairs, which rarely if ever succeed in permanent righting of the situation. On the other hand, this same experience has demonstrated that satisfactory results can be achieved if the dynamic stresses of frozen ground are carefully analyzed and are allowed for in the design in such a manner that they appreciably minimize or completely neutralize and eliminate the destructive effect of frost action. Successful adaptation of engineering principles to meet the unique permafrost conditions can be achieved only if the natural phenomena of frozen ground are thoroughly investigated and understood and their forces are correctly evaluated. A comprehensive and systematic study of frozen ground in the far North, therefore, should constitute an integral part of the planning and design of any engineering project.

The first preliminary step in the planning of any engineering project is to map the areal extent, continuity and thickness of permafrost. Such a survey, based on the examination of test-pits, drill-holes and on other known field methods, should determine:

1. Continuous areas of permafrost
2. Areas mostly underlain by permafrost but having scattered patches of belts of thawed ground ("taliks")
3. Areas of thawed ground that have sporadic, island-like masses of permafrost, and
4. Areas entirely free of permafrost

From the nature of this preliminary survey based on scattered test-pits and drill-holes, it will be apparent that the boundaries between these four types of terrain will not be precise. Even where accurately established, for certain limited stretches, these lines may in the course of years shift their position in one direction or another depending upon whether at that particular locality the permafrost condition is degrading (receding) or aggrading (spreading). Nevertheless, the practical value of such a rough classification of land is evident when it is realized that each of these terrains is distinctive enough to permit broad generalizations in the planning of projects.

For example, in terrain "a", only the passive method of construction should be adopted, i.e. all architectural specifications and designs should be such that the thermal regime of the ground is not disturbed. In such areas only the surface waters can be counted on for water supply unless the entire thickness of permafrost is drilled through to tap the ground-water below the permafrost (subpermafrost water).

Windows of unfrozen ground ("taliks") in terrain "b" are likely to serve as conduits permitting the water from both beneath and within the permafrost to percolate to the surface. For this reason the surface taliks should receive first attention in the survey for possible water supply.

In the terrains "b" and "c" either the passive or active method of construction can be used depending upon the thickness and the temperature of the permafrost. In the terrain "c", where only small islands of perma-frost are present their location can be ascertained and the projected structures can be erected in the adjacent areas which are free of permafr

It is increasingly evident that in the past the choice of many airfie sites, in the permafrost area, at which difficulties in construction and maintenance are now being experienced, has been primarily guided by the small amount of ground to be moved during the initial grading and levellir operations with little or no consideration of more critical factors which should determine the suitability of a site. It therefore follows that a selection of a satisfactory site should be given more careful attention.

For any engineering project the preliminary permafrost survey should consist of the following three stages:

1. Reconnaissance survey
2. Preliminary survey
3. Final survey

Reconnaissance survey. A reconnaissance survey consists of examin-ing the terrain from the air and of selecting of several tentative sites. These sites are then photographed and studied on the ground. Each site is evaluated as to its general adequacy and in regard to the availability of water supply, the search for which is carried on by a hydrologist during the reconnaissance survey, or even prior to it. A tentative plan of water mains is considered at this time and attention is given to general drainag problems and to sewage disposal.

The reconnaissance survey should be started in the spring before the break-up of the ice in rivers, before the melting of the ice in icings, a while the frost-mounds are still at their maximum. The ground parties generally carry on their work through the summer into the fall.

Other conditions being equal, preference is given to construction sites which are:

1. flat and dry
2. not subject to floods
3. free of ground-water seepages which freeze in the winter
4. free of swamps, frost-mounds, and cave-in lakes
5. entirely free of, or with only a thin layer of moss
6. distant from mountain or piedmont slopes. Areas near the foot of the break in slope are generally more likely to have ground-

water seepages, ground ice (lenses and wedges), and are subject to intensive frost heaving.
7. distant from canyons and gullies.

Preliminary survey is carried out to select the most satisfactory site of the several that were provisionally chosen during the reconnaissance study. Preliminary surveys should be carried out during the spring, summer, and early fall and should aim to obtain the following information:

1. Topographic map (with contour interval not more than 5 feet) of the site and vicinity to give a picture of the behavior of surface waters and possibility of drainage.
2. Location on this map of proposed water mains and sewage pipes, with profiles along these lines.
3. Distribution of vegetation (forest, shrubs, grass-land, swamps, etc.) the following should be indicated on the map:

   a. Dominant kind of trees and their timber quality.
   b. Detailed record of grasses.
   c. Character and thickness of moss and peat.

4. Character of the ground profile or cross-section along several directions across the site, texture of the ground, etc.
5. This study is carried out by means of test drill-holes and test pits and by the laboratory study of samples. Ground temperatures are read at a given time in some 15 or 20 holes situated in different environments.
6. In permafrost areas the following conditions are regarded as necessary to insure stability of buildings:

   a. A considerable thickness of permafrost - not less than 75 feet.
   b. A more or less constant sub-zero temperature from a depth of 30 feet downward. Temperature should not be higher than - 1.0° C. (30° F.).
   c. Permafrost free of moisture (Dry Permafrost) or permafrost with moisture (ice) content not exceeding saturation.
   d. Homogeneous and evenly-thawing active layer where this layer does not exceed 5 or 10 feet.
   e. An active layer that does not swell - sand or coarse aggregate containing less than 3% of fines (grains smaller than .07 mm).

7. Areas with layered permafrost are unsuitable and should be avoided. However, if no other ground is available a thorough investigation of conditions should be made before a structure is erected. Also unsuitable are areas with seepages of groundwater (which freezes in the winter), areas with frost-mounds, and where permafrost contains layers of ice not far beneath the floor of a projected excavation.

<u>Final survey</u> should begin after the completion of preliminary studies and should be carried out in the fall and through the winter. Under favorable conditions this investigation may be completed by August or September of the following year. By this time the survey of water supply should be completed.

1. It is desirable that the person who is to be in charge of the completed project or who is concerned with the design of the project, and who is, therefore, well informed on the function and behavior of the structure, should participate and cooperate in this final investigation.
2. The purpose of the final investigation is to supplement the information already obtained during the preliminary survey so as to furnish all the necessary data for planning and computation of specifications of the projected structure. Thus the final survey calls for:

   a. More detailed topographic map of a more restricted area - the actual site - with the contour interval of about two (2) feet.
   b. More detailed information on the ground-water regime. Additional test holes may be necessary. These holes should be drilled between the middle of September and October - the time of maximum seasonal thawing. Particular attention should be paid to holes with artesian flow and to the presence, amount, and the direction of percolation of ground-water.
   c. Moisture content of the grounds (in the form of water or ice) should be determined for the active layer and for the upper part of the permafrost. These data are necessary for computing the adfreezing strength. The amount of ice in the permafrost, expressed in percentage, will be used as a basis for various computations including the amount of expected settling of ground when thawed, and the amount of stored cold.
   d. More detailed record of thermal regime of ground.
   e. More accurate data on permafrost table.
   f. Detailed analysis of the nature and amount of swelling of the active layer.
   g. Laboratory tests of physical properties of soils.

3. In October, after all these observations have been completed, the building site is stripped of vegetation and sod, and is appropriately graded. During the winter the snow should be kept off the surface and in August or September of the following year all observations on the depth of permafrost table should be repeated and additional test pits examined. The observed change in the level of the permafrost table should be accurately recorded.
4. Timing of the surveys is shown graphically in the following diagram.

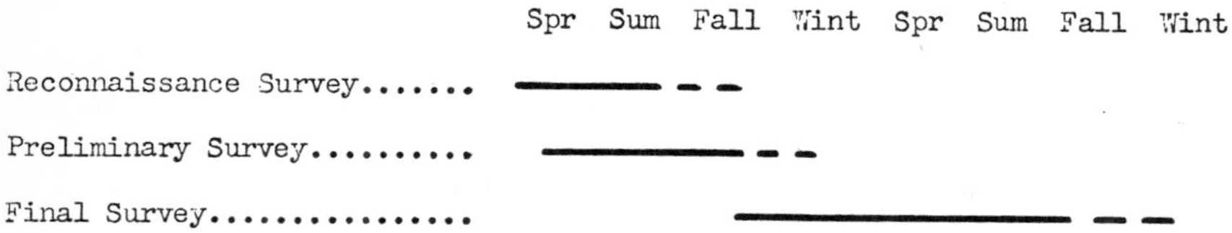

	Spr	Sum	Fall	Wint	Spr	Sum	Fall	Wint
Reconnaissance Survey.......								
Preliminary Survey..........								
Final Survey................								

It should be repeated again that the completion of comprehensive permafrost surveys may require as much as a yoar or more. Under the pressure of the existing war conditions this may be impracticable, but, notwithstanding this shortcoming, it is deemed necessary to present the recommended outlines in full for they are intended to serve as check-lists, leaving it to the individual executive engineer to modify the procedure and to adopt the plan appropriate to each problem. It should be borne in mind, however, that the taking of shortcuts in the recommended procedure might result in leaving out of consideration factors that are critical to a particular problem with the possibility of subsequent partial or complete failure of the project.

Geophysical Methods in the Study of Permafrost

During the past ten years some progress has been made in the application of the electrical resistivity method to the study of permafrost. This method appears to be most effective in areas where permafrost is not very thick (of the order of 60 meters) and where its temperature is not lower than -1.5° C. For areas where permafrost is thicker and its annual temperature is low only qualitative results can be obtained through the resistivity methods.

Theoretically, ice acts as an insulator whereas the unfrozen ground above and below has only a small specific resistivity. (Fig. 80). The resistivity of frozen grounds depends on their composition, porosity, content of moisture, concentration of dissolved salts, and temperature. For example, the resistivity of sandy clay with 100% of moisture and at the temperature of -6.4° C. is 20,000 M^2/M. Sand with 100% moisture at the temperature of -8° C. has the resistivity of 33,000 M^2/M. The commonest type of ground met in the field with a temperature between 0° and -2° C. will have, on the average, the resistivity of about 1,600 M^2/M.

In a certain area in Transbaikalia the specific resistivity of unfrozen active layer measured 50 M^2/M whereas in the underlying frozen ground, 60 meters thick at the temperature of -1.5° C. the resistivity was from 1,000 to 2,000 M^2/M. The resistivity of the unfrozen ground (talik) below the permafrost was approximately the same as that of the surface unfrozen ground.

In the extreme north, on the Taimyr Peninsula, the temperature of the permafrost reaches -12° C. and its thickness is in the order of

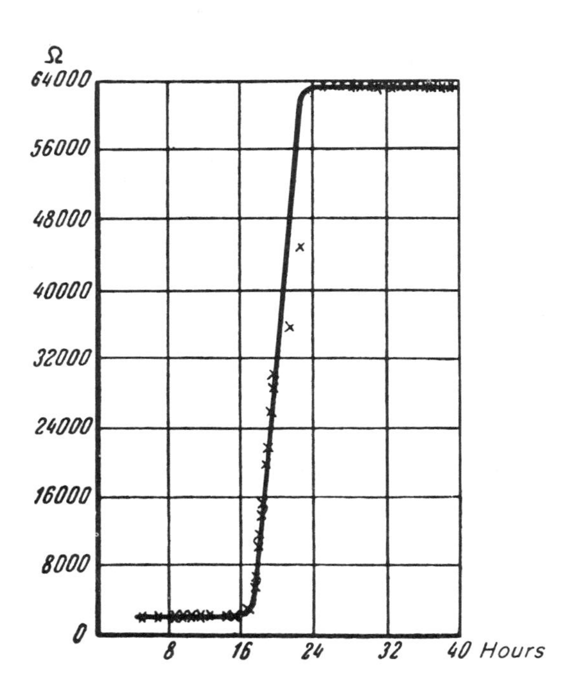

Electric resistivity of ground during the process of freezing.

(From Tsytovich and Sumgin)

FIG.80

350 meters. There the resistivity of frozen ground is as much as 50,000 M^2/M.

As the conductivity of grounds is electrolitic rather than metallic it follows that the lower the temperature of the ground the greater amount of the electrolite will pass into the solid state thus increasing the specific resistivity of the ground. On the other hand, the higher the temperature of the ground the greater mobility is possessed by the electrolite and the resistivity of the ground is considerably less.

Where the permafrost is very thick and its temperature is very low it contains only a very insignificant amount of the liquid electrolite in disconnected particles. The passage of a direct current through such a mass is very difficult and subject to various distortions which confuse the interpretation of results.

Basically the application of a direct current to the study of frozen grounds consists in passing the current into the ground at two distant points A and B, and in reading the voltmeter connected with the ground at two points M and N which are placed along the line AB equidistant from a central point O. The obtained reading applies to the point O', situated below the point O at a depth variously computed by Schlumberger, Petrovsky and others as equal to $\frac{AB}{4}$, $\frac{3AB}{8}$ or $\frac{AB}{3}$. By moving the set-up from place to place and by varying the distances between A and B the desired results may be achieved as to the horizontal and vertical distribution of the frozen condition of the ground.

A variant of this method with vertically suspended wires may be applied to the survey of a frozen ground in a drill hole. In recent years, seismic and other geophysical methods have been employed in the survey of the permafrost but the information on the methods employed and results achieved is as yet too meagre to be of value.

Test Pits and Drill Holes

Test pits give the most satisfactory results in revealing the character of the active layer, of the permafrost and of the regime of ice and water. Unfortunately, it is difficult to sink pits of 1 x 1.5 or 1 x 2 meters dimensions in the frozen ground. These difficulties are intensified by the presence of water above the permafrost for, in spite of all the precautions such as structural reinforcements and lining of pits, the hole frequently becomes flooded and its walls slump and flow. The work is unavoidably very slow.

It is, therefore, very important to choose carefully the sites for test pits and to distribute them over the area in such a way that the maximum amount of information may be obtained from as few test pits and drill-holes as practicable.

Test pits are usually dug to the depth of 5 meters and greater depths are explored by drill holes. Notwithstanding the difficulties, a few deep test pits should be excavated at critical places on the site that is surveyed for a major structure.

Two and three inch drills are commonly used in connection with road and railroad surveys. But for buildings and industrial plants, where the equipment does not have to be moved from place to place over a considerable distance, a larger diameter drill is recommended.

In passing through the active layer it is necessary to use casing regardless of the nature of the ground. On the other hand, in drilling through permafrost that is not interlayered with taliks, casing may not be necessary.

The casing in the active layer should be extended downward beyond the base of the active layer and should penetrate the permafrost to a minimum depth of 1.5 meters.

The taliks in the permafrost should be sealed off by casing and those taliks which are encountered at the bottom of the hole should be plugged. An effective plug consists of alternating layers of clay and oil-soaked burlap.

The use of drilling mud is permissible only when no observations on temperature of the ground and amount of ice are required. Core samples should be taken at frequent intervals and should be recorded together with the information on the relative resistance of ground to penetration. Samples are usually taken with every change of ground and in a homogeneous material at intervals not less than .5 meter.

Samples are carefully studied and described as to their lithology, texture and other physical properties. Ice inclusions are also carefully examined, measured and recorded. Samples are then thoroughly dried, packed, and labelled indicating the specific purpose for which they are obtained so that appropriate tests may not be overlooked during the laboratory examination.

The spacing of test pits and drill holes is governed by the size of the building site, the type of projected structure, and by the character and the degree of uniformity of the ground. It stands to reason that in an area of undisturbed alluvium of uniform lithology there is no need to dig as many test holes, as where conditions are markedly variable.

In surveying for a highway or railroad it is recommended that the test holes be located at the following places:

1. On the slopes of different exposure.
2. On the slopes of different declivity.
3. In areas of different soil, vegetation, and minor features of relief.

4. In places of projected excavations for foundations or roadcuts.
5. In places of projected fills.
6. In swampy hollows and depressions.
7. At the sites of springs and icings.
8. At cave-in lakes.
9. Near landslides and slumps.
10. In areas of ground-ice.
11. At stream crossings.
12. At sites of heavy structures.
13. At the sites of pits and quarries for building material and for ballast.

Under the most uniform conditions test pits may be spaced 250 - 500 meters apart along a projected road, but where the conditions are variable test pits should be not farther than 50 meters apart.

Where a projected road will pass over a bedrock a few shallow pits to determine the extent of weathering will suffice. Sites that are selected for a station building or a platform should be tested at intervals of from 20 to 50 meters.

Where a project calls for no excavation work and only for a moderate amount of fill it will suffice to sink test pits to the depth of three times the thickness of active layer or even less if bedrock is encountered before that depth is reached. The places of projected excavation should be tested to a depth of from 1 to 2 meters below the ultimate level of the excavation.

With the active method of construction - where the permafrost condition is eliminated - test pits should penetrate to a depth of about 5 times the width of the foundation of a projected structure.

In all explorations at least two deep test drills penetrating below the level of zero annual amplitude should be made in order to ascertain the temperature gradient of permafrost. The temperature gradient and the thickness of the permafrost are the basic factors which guide the choice of either the passive or active method of construction.

A complete description of the exposed wall of a test-pit should be made. If possible the inclination (attitute) and extent of layers should also be recorded. Particular attention should be given ice inclusions, their character and structure. The walls of a pit should be sketched and, if possible, photographed. As core samples from drill holes are not large enough to make certain tests, the samples for such tests are usually taken from test pits. Two sizes of samples should be taken. A small sample for general study and a sample of about 10 kg. for mechanical tests. If the active layer is homogeneous throughout its entire thickness the following minimum number of samples will be sufficient.

1. From the middle of the active layer.
2. At the contact with permafrost.

3. From the upper part of permafrost - at the base of the projected foundation, and
4. From a depth of 2 to 4 meters below the base of the projected foundation.

The thickness of the active layer and the depth to the permafrost table should be determined in test pits as well as in drill holes.

The thickness of active layer at the time of excavation usually does not give its true (maximum) value. An approximate measurement can be obtained late in the autumn but the true value is determinable only after several years of observation.

But even though a single observation during the exploration does not give the true value of the thickness of active layer, it nevertheless provides valuable comparative data which upon a correction based on the autumn measurements may be brought to a close approximation of the true thickness. These corrections are based on the fact that the summer measurements are made when the ground has not yet thawed to the maximum depth. The first correction is therefore made to obtain the maximum thickness for that particular year. If that maximum should prove to be the greatest recorded over a period of many years then no other correction is needed. If, however, the greatest depth of thawed ground for that year is less than the maximum measurement for some other year, then a second correction is added.

Another method of determining the true thickness of active layer is based on a careful study of the vertical profile of the ground with respect to moisture, lithology and organic content. The ground immediately above the permafrost table commonly contains a relatively greater amount of moisture and has a considerable concentration of mineral salts. The relative proportion of fine clastic particles of less than .005 mm is also noticeably higher near the bottom of the active layer. This point, at least in part, may be illustrated by the following graph (Fig. 81) based on Guterman's work in the Lower Yenisei region.

Determination of Moisture Content

A rapid and most convenient method of determining the moisture content is the calculation of moisture by weight. Samples in test pits are taken at vertical intervals not greater than .5 meter and also at every change in the character of the ground. The following procedure is recommended:

1. The face of the cut is thoroughly cleaned.
2. A steel cylinder of known volume is pressed into the ground, turned several times, and pulled out with the sample.
3. The sample is pushed out into a clean and dry container of known weight.
4. The sample is weighed before and after the moisture is driven off

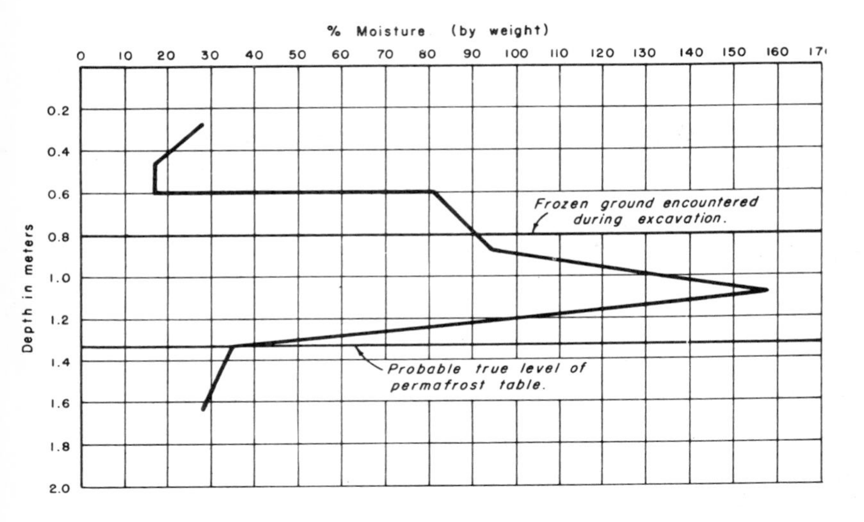

Moisture profile of frozen ground indicating the level of permafrost table.

(After Guterman)

FIG. 81

In gravelly and bouldery ground where the cylinder cannot be driven into the face of the cut, the cylinder is filled with chipped fragments which are carefully selected to give a representative sample.

Determination of moisture content in a drill sample is more painstaking and is not always reliable. The presence of ground particles from higher levels and of water at the bottom of a hole make the results inaccurate.

The following procedure tends to minimize the error:

1. A few centimeters before the bit reaches the ground which it is planned to test, the hole is thoroughly cleaned and is pumped dry.
2. Drilling proceeds to the level at which the ground is to be tested.
3. Two samples are taken and placed in tightly capped containers.
4. The usual procedure is then followed.

Measuring of Ground Temperatures

Uses of Ground Temperatures. Taking of Ground Temperatures serves the following purposes:

1. Establishes the thickness of the active layer.
2. Establishes the presence of taliks.
3. Determines the depth of seasonal fluctuation of ground temperatures.
4. Determines the temperature profile of the permafrost.
5. Determines the areal extent of the permafrost, areas of continuous permafrost, variations in the permafrost, etc.
6. Establishes the effect of buildings and other structures on permafrost, showing whether permafrost is receding or increasing.
7. Determines the relative heat conductivity or relative insulating value of the different kinds of ground and building materials.
8. Furnishes the necessary data in prospecting for water supply.
9. Furnishes the necessary data to determine whether a passive or active method of construction is to be adopted. (With the passive method the permafrost is preserved or is left undisturbed. With the active method the frozen ground is thawed out before the structure is erected.)
10. Provides necessary data for computation of building specifications, necessary amount and kind of backfill, required insulation for buildings, sewer and water lines, etc.
11. Enables one to determine whether or not piles driven into a pre-thawed ground are completely frozen back and are ready to be loaded with the computed weight of the structure.
12. The time data obtained from the above (#11) may be used in planning the schedule of construction of new projects.
13. Permits interpolation and extrapolation of data for intermediate and nearby localities.
14. Provides the data for necessary corrections of single readings in the nearby areas.

Ground Temperatures may be measured with:

1. Glass thermometers.
2. Thermal bulb thermometers with extensions.
3. Thermocouple thermometers, or
4. Resistance thermometers.

Glass Thermometers. Ordinary glass thermometers are simple to operate but are likely to be inaccurate and do not permit reading of temperatures at a distance. The so-called SLOW-RECORDING thermometers have the advantage that they do not change their reading while being lifted from a test hole through the air that has a different temperature

The bulb in a slow-recording thermometer is inclosed in a hard rubber, wax, or other insulating material to slow down its reaction to the changes of temperature making it possible to read temperatures in a drill hole down to the depth of about five hundred feet (500').

A slow-recording thermometer may be constructed by placing an ordinary glass thermometer into a waterproof metal case with packing of insulating material around its bulb.

Under certain conditions, especially in taking temperatures of shallow excavations, maximum and minimum thermometers may be used. Maximum thermometers can be used in the winter when ground temperatures are higher than those of the air and minimum thermometers can be used in the summer when the reverse is true.

Thermal bulb thermometers with extension consist essentially of a a temperature sensitive bulb, connecting tubing, hollow elastic element or Bourdon tube, compensating element, and Pen-Arm.

The entire system, from bulb to the elastic element, is filled with vapor, inert gas or liquid. Changing temperatures at the bulb result in changes in pressure (or volume of liquid) in the system. These changes actuate the elastic element, which through pen-arm records the changes in terms of temperature.

Vapor pressure type of thermometers, known in the trade as Class II, register temperatures ranging from -25 to +560 F°. The bulb is small and can be supplied with tubing extension up to five hundred feet (500').

Gas pressure thermometers (Class III) can register temperatures from -100 to +1000 F°. They can be equipped with tubing extension up to three hundred feet (300'). Liquid expansion thermometers, known as Class I and Class IV, have a range from -150 to 500 F°. Class I can be used only when the case and tubing are at the same temperature. This feature makes them unsuitable for permafrost studies. On the other hand, Class IV thermometers can be used where temperatures may differ along the tubing. These thermometers appear to be especially useful for measuring of ground temperatures. They may be furnished with tubing up to one hundred fifty feet (150') long.

Thermocouple Thermometers. The thermocouple works on the principle that when two (2) wires of dissimilar metal are joined at each end to form an electrical circuit, an electrical potential proportional to the temperature difference existing between the two (2) junctions is generated.

The potential will then cause a current to flow, the strength of which is dependent upon the total electircal resistance of the circuit. This current may then be used to operate an electrical instrument for indicating temperature.

The measurement of temperature by means of thermocouple thermometers involves a measurement of the millivolt output of the thermocouple. A sensitive millivolt meter is sometimes used but the limitations of this device often necessitate the use of a potentiometer, which measures the voltage output without requiring a current to flow in the thermocouple circuit.

The thermocouple thermometer consists of a thermocouple, leads, and indicator.

The indicator is a senstitive mechanism having a coil arranged to move in a annular air gap of a permanent magnet. The coil that carries the pointer and the control is two (2) phosphor bronze springs which also serve to conduct the current into the coil.

The thermocouple leads and thermocouple are usually made of iron and constantan to form a couple, one wire being iron, the other constantan Constantan is an alloy of copper and nickel.

The thermocouple thermometers are manufactured with carefully calibrated lengths of the leads and for this reason their use is generally limited to holes which do not exceed 30 or 50 feet in depth.

Thermocouple thermometers are particularly suitable for a permanent installation where daily temperatures are to be read. Where several thermocouples are installed in a single pit or a drill hole, a transfer switch may be used making it possible to take as many as twenty (20) different readings in rapid succession.

The advantage of thermocouple thermometers over other instruments are:

1. Thermocouple thermometers can be placed in locations where ordinary thermometers cannot.
2. As many as twenty (20) different readings can be taken in rapid succession.
3. Possesses a high degree of accuracy.

It is important to remember that the thermocouple leads must not be lengthened or shortened as they are of a definite resistance and enter into the calibration of the indicator. For this reason utmost care should be exercised during the lowering and pulling of the couple to keep the strain on the cord and not on the leads of the couple.

Resistance Thermometers. The resistance-type thermometer consists of three (3) main parts: the temperature sensitive element (bulb), indicator, and the connecting lead wires between the indicator and the bulb.

Temperature measurement by the resistance-type thermometer is based on the use of a suitable coil of wire, usually nickel or platinum, the electrical resistance of which varies with its temperature. The resistance thermometers are highly accurate and sensitive. They are especially suited for distant reading installations in deep pits and drill holes. The only disadvantage of the resistance thermometer is its high cost.

<u>Location of Test Pits.</u> It is very important to choose the sites for test pits and to distribute them over an area in such a way that the maximum amount of information may be obtained from as few test pits as practicable.

In a general permafrost survey, ground temperatures should be measured at various depths below the surface in all distinctively different landscapes. Thus it will be necessary to locate pits in open areas, wooded areas, in sandy ground, clayey or silty ground, under paving blocks, in areas of heavy snow cover, in areas with little or no snow, and so on.

It is also recommended that one or several buildings in a typical permafrost environment be chosen for a careful study of the frozen ground beneath them. This record of ground temperatures should be accompanied by a detailed description of the design and materials used in these structures and the amount of heat generated in their various parts by stoves, boilers, furnaces, etc.

In surveying for a highway, railroad or a large building an effort should be made to obtain the data on ground temperature (and other characteristics of frozen ground) at the following critical locations:

1. Sites of proposed buildings, roads, runways, etc.
2. Along projected utility lines (water, sewerage, steam, etc.)
3. At sites of projected excavations for foundations, roadcuts, drainage ditches, etc.
4. At sites of pits and quarries for building material, backfill, or ballast.
5. At sites of projected fills.
6. In areas of different soil, vegetation, and minor features of relief (hummocks, hillocks, "nigger-heads", frost-mounds, etc.)
7. On slopes of different exposure.
8. On slopes of different steepness (declivity).
9. In swampy hollows and other depressions.
10. Near ground-water seepages.
11. At lake shores and along streams (at crossings) near the site of a project.
12. In areas where ground-ice is known or is suspected to exist.
13. Near landslides and slumps.

Test pits for ground temperatures (and for other observations of frozen ground conditions) are rarely excavated to a depth of more than fifteen (15) feet (five (5) meters). Drill holes are generally used to explore the ground at greater depths. However, notwithstanding the difficulties which are encountered in the excavation of pits, a few deep test pits should be sunk at critical places on a site that is being surveyed for a major structure.

In general, the spacing of test pits is governed by the size of the building site, the type of projected structure, and by the character and the degree of uniformity of the ground. For instance, in an area of undisturbed alluvium of uniform composition, there will be no need to dig as many test holes, as where conditions are markedly variable from place to place.

Under the most uniform conditions test pits may be spaced five hundred (500) to two thousand (2,000) feet apart but where the conditions are variable test pits should not be farther than two hundred feet (200') from each other. On building sites test pits are generally spaced from fifty to two hundred feet apart.

In a bedrock area a few shallow pits through the weathered (decayed and disintegrated) surficial material will suffice.

Excavation of Pits. Test pits for temperature observations should be dug without preliminary thawing of the ground. Test pits are dug to depths of several feet down to not more than fifteen feet (15'). Exploration of ground at greater depth is usually carried out by means of drill holes. However, at least one (1) deep test pit should be sunk at each major project.

Excavation of frozen ground without preliminary thawing is admittedly a very difficult and unavoidably slow task, especially if the ground is clayey or silty. The difficulties are intensified by the presence of water above the permafrost and in spite of all the precautions (structural reinforcement, plank lining, etc.) the hole frequently becomes flooded and its walls slump and flow.

The use of explosives has not proved to be very effective although it is worth noting that of the different explosives tried by the Russians in blasting frozen ground, the best results were obtained with "Ammonite" or "Ammonal". This explosive does not require thawing and is safe to transport. Its disadvantage is that it is hygroscopic and therefore absorbs moisture. This, however, can be overcome through the use of waterproof packing or wrapping.

To prevent the slumping and sloughing of the walls, the active layer at the mouth of the hole should be cased off. The casing may be made of closely fitting, sturdy plank cribbing and the space on the outside of the crib should be tightly packed with a tar or oil-soaked burlap interlayered with clay or thick mud and capped at the top with moss, peat, or other insulating material. The casing should extend downward through the entire thickness of the active layer and should continue into permafrost at least to a depth of three (3) or four (4) feet.

For measuring ground temperatures in a pit that is being dug, the excavation is carried out to a depth seven inches (7") short of the level at which it is planned to measure the temperature. Using a wooden guide (Figure 82), a hole ten inches (10") in length is bored at a 45° incline at the bottom of the north-facing wall. The end of the thermometer placed in this hole will thus be at an even foot or the level of the bottom of the hole plus seven inches (7"). The purpose of boring

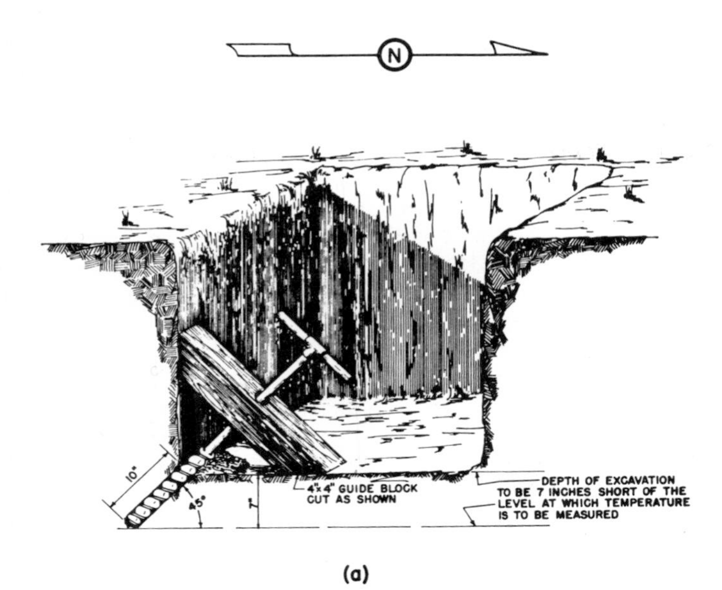

(a)

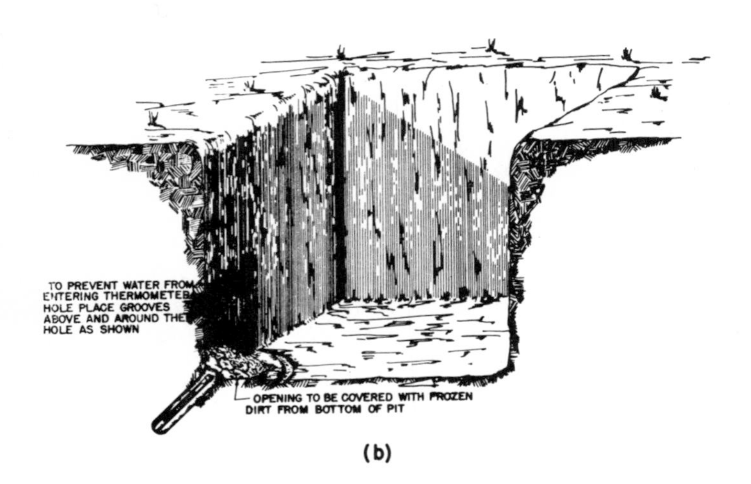

(b)

MEASURING GROUND TEMPERATURES DURING EXCAVATION

FIG. 82

the hole at the 45° incline is to reach the ground the temperature of which is least affected by the temperature of the air that may penetrate into the ground through the exposed walls and the trampled bottom of the pit.

In sloping ground, the distance from the surface is measured normal (at right angle) to the plane of slope and not to the horizontal plane.

If the ground is wet and the water or mud drips along the sides of the pit, a groove is made on the wall above the thermometer to prevent the penetration of water into the thermometer hole. A similar groove may be necessary on the bottom of the pit to keep the bottom moisture out of the hole. (Figure 82). Test pits should be large enough, not less than 3' x 3', to provide the recorder with ample "elbow room" to manipulate the instruments.

Where a project calls for no excavation and only for a moderate amount of fill, it will suffice to sink test pits to the depth of three (3) times the thickness of active layer or even less if bedrock is encountered before that depth is reached. The places of projected excavation should be tested to a depth of five (5) to ten (10) feet below the ultimate level of excavation.

Where the active method of construction is being tentatively considered, test pits should penetrate to a depth of about five (5) times the width of the foundation of a projected structure.

Test pits excavated for periodic recording of temperature should be covered with a well-insulated, double lid. It is also recommended that the walls of the pit be lined with planking down to the very bottom, with the space between the planks and the ground packed with moss, peat or other insulating material. Ground temperatures in such a pit should be read in a side bore hole drilled horizontally. The bore holes should be at least six feet (6') long, preferably longer. The openings of these side bore holes should be kept plugged at all times except, of course, during the insertion of the thermometer.

During the excavation of a pit the exposed wall (profile) should be described in detail, sketched, and, if possible, photographed. The inclination (attitude) and the thickness and extent of individual layers of ground should be studied, measured and accurately recorded. Particular attention should be given ice inclusions, their character, structure, and distribution.

Samples for soil analysis and for mechanical test of ground should be obtained during the excavation of a pit. Two (2) sizes of samples should be taken. A small sample to fill a quart jar for general study and a large sample, about twenty (20) pounds (ten (10) kg) for mechanical tests. If the active layer exposed in the pit is homogeneous throughout its entire thickness the following minimum number of samples is generally taken:

1. From the middle of the active layer.
2. At the base of the active layer, near the contact with permafrost.
3. From the upper part of permafrost - at the base of the projected foundation, and
4. From a depth of six (6) to fifteen (15) feet below the base of the projected foundation.

The most suitable time for excavation work is during the months of September, October, November, March, April and the first half of May. During the summer the ground is too wet. Little or no work is carried out during the winter due to the adverse or intolerable weather.

Measurement of Temperatures in Pits. In test pits ground temperatures are usually measured every foot or two feet. Where the distance to permafrost is only about one foot (1'), an intermediate measurement of temperature should be made at about six inches (6") below the surface of the ground. Intermediate measurements should also be made between any two (2) points if the temperature in that interval shows a sudden change.

The temperature of the air and of the surface of the ground should also be measured. These measurements may be made with an ordinary thermometer, but beginning with the depth of one (1) or two (2) feet only insulated, slow-recording thermometers should be used.

For measuring the temperature of the surface of the ground, a slight depression is made to accommodate the thermometer in a horizontal position in such a way that it will lie parallel to the surface of the ground. The thermometer is then covered with about one-eighth (1/8) of an inch of dirt scraped off the ground nearby and is left there for ten (10) or fifteen (15) minutes.

For measuring the temperature at six inches (6") below the surface, a hole may be bored to that depth with an auger, a thermometer is placed in the hole and is covered with sod or moss. The reading is taken in about ten (10) or fifteen (15) minutes.

Measurement of temperatures in pits that are being excavated is made in ten inch (10") inclined drill holes at the bottom of the north-facing wall (Figure 82). After the thermometer is placed into this short bore hole, it is covered with frozen ground obtained from the bottom of the pit. (Figure 82).

The thermometer is kept in the hole from ten (10) to thirty (30) minutes depending upon the extent to which the temperature of the ground has been disturbed (changed) during the drilling. Usually little, if any, heat is generated in soft ground but in a hard ground the frictional heat during the drilling may be considerable, thus a longer time will be required before the true temperature of the ground is restored.

The thermometer should be removed from the hole without jerky movements so as to avoid the "slipping" or "breaking" of mercury in the glass tube.

In taking periodic readings of temperature in a pit with long side bore holes, the thermometer may be shoved gently into the hole with a slender wooden rod or a stiff wire fitted with several disc-shaped cloth or leather "baffles" frilled around the edges and loosely fitting into the hole. These "baffles" will prevent or minimize the convection of the air in the hole so that the temperature at the end of the hole will be as nearly as possible the same as in the immediately surrounding ground. (Figure 83).

As a rule it is sufficient to make only one (1) reading, but if there is a reason to doubt the correctness of the reading, as may be the case if the temperature shows a sudden jump from the previous measurement, several readings should be taken at intervals until the recorder is satisfied that the reading represents the true temperature of the ground.

To prevent the freezing of a thermometer to the wall of a pit or of a side bore hole it should be wiped dry and coated with a thin film of grease. A single layer of paper wrapping and a coat of grease may be used, but care should be exercised not to obstruct the view of the graduated side of the thermometer. In test pits the thermometer should be turned around every five (5) minutes.

The record of ground temperatures (see table on page 182) should be accompanied with a general statement on weather conditions - whether clear, cloudy or rainy. In the winter the amount of snow and its compactness should be carefully examined, measured and recorded.

Location of drill holes. Selection of sites for test drill holes is governed by the same rules and considerations as in the case of test pits. They should be chosen in such a way that the maximum amount of information may be obtained from as few drill holes as practicable.

In all explorations at least two (2) deep test drills, penetrating below the level of zero amplitude, should be made in order to ascertain the temperature profile (gradient) or permafrost. The temperature gradient and the thickness of the permafrost are the basic factors which guide the choice of either the passive or active method of construction.

Wherever feasible at least two (2) holes should be sunk within one hundred (100) or two hundred (200) feet of each other on similar or identical slope, exposure, soil, and vegetation in order to verify the general range of temperature.

Drilling of test holes. Test holes for measuring ground temperatures should be drilled without preliminary thawing of the ground.

Shallow test holes down to fifteen feet (15') may be drilled by hand with an auger or with a post hole digger.

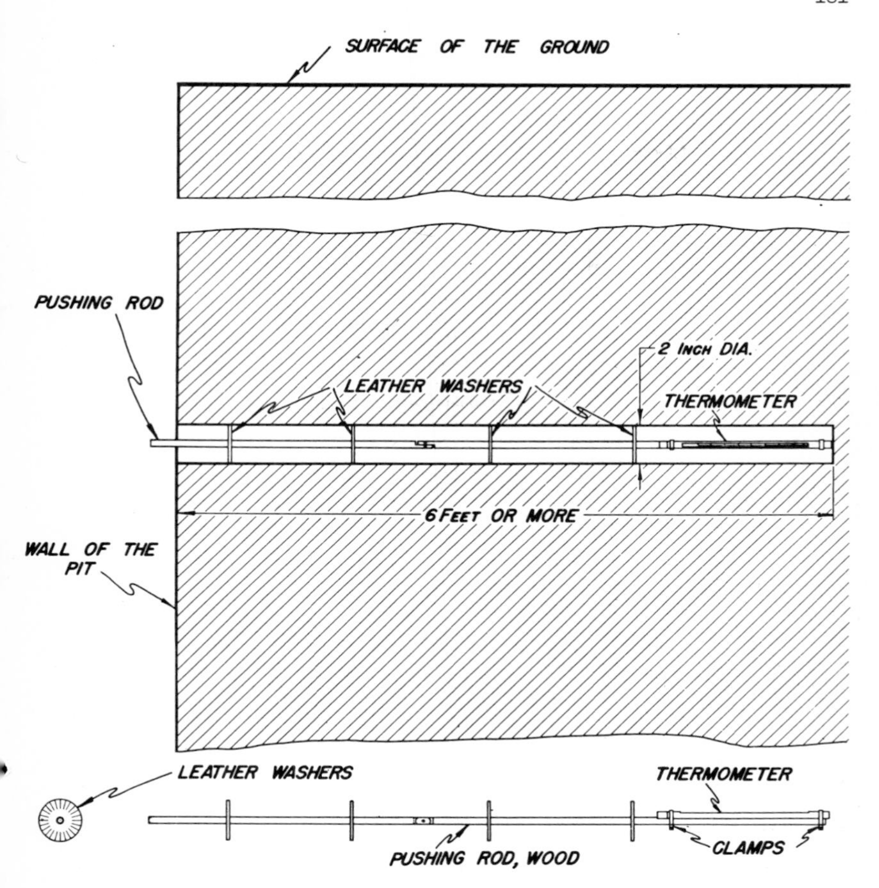

MEASURING TEMPERATURES
IN A SIDE BOREHOLE

FIG. 83

RECORDING OF TEMPERATURES

The following forms may be used to record ground temperatures:

Glass and Thermal Bulb Thermometers

TEMPERATURE RECORD

LOCATION:

Hole or Pit No: Depth to Permafrost Table:

Date	Hour (1-24)	Min.	Depth in Feet	T° Reading in C°	Corrected T° to C°	Conditions: Clear, Cldy. Rain, etc.	Remarks	Recorder
			Air					
			Surface					
			of gr.					

Thermocouple Thermometer

TEMPERATURE RECORD

LOCATION:

Hole or Pit No: Depth to Permafrost Table:

Date	Hour (1-24)	Min.	Depth in Feet	Reading of the Control Thermometer	Reading of Galvanometer	T° in C°	Remarks	Recorder
			Air					
			Surface					
			of gr.					

Resistance-Type Thermometer

TEMPERATURE RECORD

LOCATION:

Hole or Pit No: Depth to Permafrost Table:

Date	Hour (1-24)	Min.	Depth in Feet	Ohms	T° in C°	Remarks	Recorder
			Air				
			Surface				
			of gr.				

Two (2) and three (3) inch mechanical drills are commonly used in connection with road and railroad surveys. For buildings, industrial plants, and landing fields, where the equipment does not have to be moved from place to place over a considerable distance, a larger tool is recommended. Description of mechanical drills and instructions for their operation are now being assembled and will be included in this manual when sufficient information is obtained.

The active layer near the mouth of the hole and water-bearing layers should be cased off by using a pipe of a slightly larger diameter than the drilling tool. This pipe should be driven down through the entire thickness of the active layer and should extend downward into permafrost to a depth of three (3) or four (4) feet. If any space is present between the casing pipe and the ground, it should be tightly packed with tar soaked burlap inter-layered with clay or heavy mud. In drilling through permafrost that is not interlayered with taliks, casing may not be necessary.

For periodic recordings of ground temperatures down to fifteen feet (15') or less, it is recommended that separate holes with a pipe casing be sunk to each level at which it is planned to take the readings (Figure 84). The upper ends of the pipes should be about six inches (6") below the surface of the ground and should be plugged to prevent any effect of the air temperature on the ground below. The plugged ends of the pipes are in turn covered with a well-insulated lid resting on a low wooden cribbing. (Figure 84).

Measurement of Temperature in Drill Holes. In drill holes down to six feet (6') in depth the ground temperature should be measured every two feet (2'). In the holes from six (6) to thirty (30) feet, the readings should be taken every three feet (3'), and from thirty feet (30') to one hundred fifty feet (150') - every ten (10) or fifteen (15) feet. At depths greater than one hundred fifty feet (150') temperature measurements should be made every twenty-five (25), fifty (50), or seventy-five (75) feet, depending upon the degree of the precision desired.

Deep holes drilled with the warmed mixtures (muds) should be measured for temperature only after the original temperature of the ground has been restored, figure 85. In such holes, preliminary test measurements should be made about once every ten (10) days.

Before any ground temperatures are measured in a deep drill hole, it is necessary to remove the drilling mud and to ascertain whether or not the temperature of the air in the hole is the same as the temperature of the immediately adjacent ground. Three (3) thermometers are suspended on a cord with one (1) resting near the mouth of the hole, one (1) in the middle, and one (1) near the bottom of the hole. The bottom instrument should be kept four (4) to six (6) inches from the bottom ground as there is always a danger of its freezing to the ground. After the thermometers are placed in the hole, the hole is closed with a wooden plug and covered with some moss, peat, or sod. Thermometers are kept in the hole at least six (6) hours.

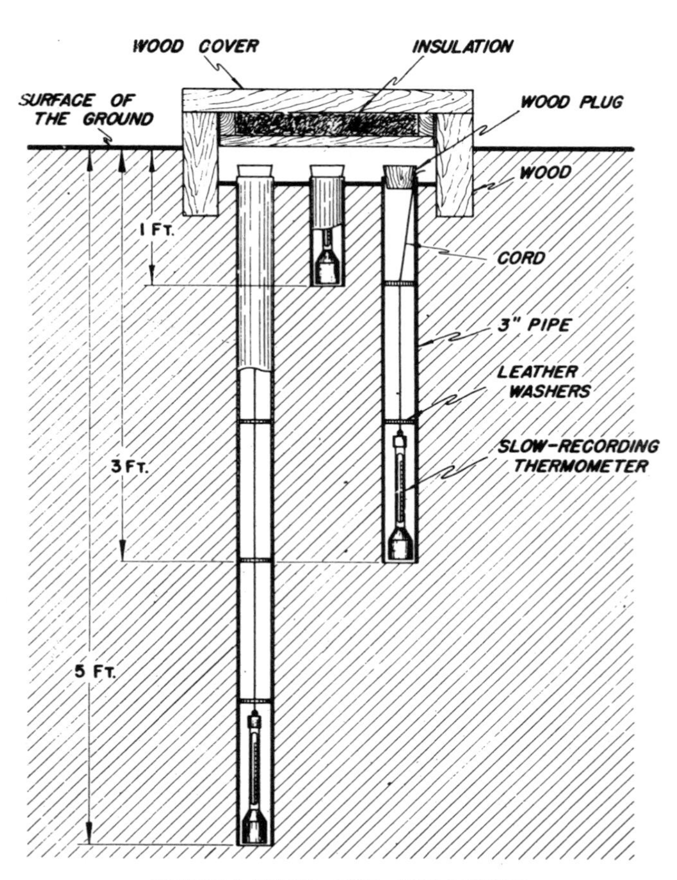

INSTALLATION FOR MEASURING GROUND TEMPERATURES

FIG. 84

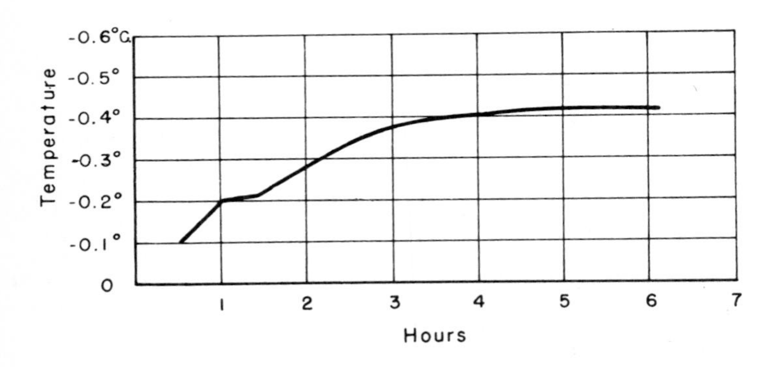

Graph illustrating restoration of temperature at the bottom of drill-hole after cessation of drilling

(From Yanovsky)

FIG. 85

At the end of the period they are pulled out rapidly but without jerking and the temperatures are read. The tenths of a degree are read first and the tens of degrees last.

These measurements are repeated three (3) times and if the results are identical (within .1 of a degree) the regular systematic measurements can be begun, otherwise the preliminary measurements should be repeated until better agreement is finally reached.

The regular measurement of temperature in a drill hole is usually made with five (5) to ten (10) thermometers which are suspended on a line at carefully measured intervals.

Slow-recording glass thermometers may be lowered into a hole on a sturdy cord fitted with one (1) or more disc-shaped, frilled-around-the-edges, cloth or leather "baffles" loosely fitting inside of the pipe. These "baffles" will prevent or minimize the air convection in the hole so that the temperature at the bottom or in any section of the hole will be as nearly as possible the same as in the immediately surrounding ground.

It is recommended that the bottom section of the hole is measured first as cave-ins and the flowage of ground are likely to make it impossible to reach the bottom of the hole even a short time after it is drilled.

Occasionally due to unavoidable cave-ins or other causes, it is impossible to remove the drilling mixture (mud) from the hole. In such cases, the temperature can be measured in the presence of this mixture, but special care should be taken to prevent the penetration of moisture into the slow-recording thermometer case, or else a special thermometer with cup attachment should be used. The small amount of liquid lifted in a receptacle will prevent the temperature indicated by the thermometer from changing while being raised and read.

To prevent the freezing of a thermometer to the wall of a drill hole, the instrument should be wiped dry and coated with a thin film of grease. The frilled edges of the "baffles", strung on the supporting cord, should also be greased lest they freeze to the pipe.

To free a thermometer that is frozen to the wall of a drill hole, a metal ring of a diameter intermediate between the width of the hole and the size of the thermometer may be gently worked down into the hole until the instrument is freed.

The thermocouple thermometer can be used to measure temperatures only in holes which are not more than fifty feet (50') deep.

The contact point of the thermocouple is tied to a cord which is weighed down with a one (1) to five (5) pound sinker. To facilitate the reading of depths in a hole, the cord should be marked every foot and with ten foot (10') special markers. In lowering and lifting the couple, care should be exercised to keep the strain on the cord and not on the leads of the couple. To avoid entangling of the wire with the cord, they should be wound on separate spools.

In measuring temperatures of a drill hole with a resistance-type thermometer, the cable should be weighted with a well-insulated sinker from five (5) to ten (10) pounds. To facilitate the reading of the depths, the suspended cable may be marked off every five (5) or ten (10) feet.

Survey of Icings

Icings may be formed by river water, by ground-water or by a combination of the two.

The study of icings may be carried out along the following plan:

The icing is numbered, plotted on the map and the following data are recorded in the notebook:

1. Name, source (whether river or spring).
2. Geographic location, altitude (how determined), also height above the nearest river.
3. Situation of the icing in the river valley, characteristic features of the river and valley.
4. The name of the drainage basin (large river).
5. Distance and direction to the nearest settlement.
6. Position of the icing with reference to the surrounding relief (whether in the bottom of the valley, at the foot of the slope, on the slope, on a terrace, on an alluvial fan, or on a saddle of a divide, etc.).
7. Position of the icing with reference to the compass orientation of the features of relief, whether facing south, north, or east.
8. Geology and hydrogeology of the icing and of the surrounding area.
9. Depth, thickness, and the character of the active layer and of the permafrost.
10. Variations in the thickness of the ground above the permafrost in different directions from the source of the icing.
11. Direction of subsurface percolation or flow of water.
12. The type of the icing, whether river, ground or mixed.

   A. For an icing formed by ground water:
      1. Whether fed by the surficial water or by the water from below the permafrost.
      2. Examine water-bearing layers or water-filled fissures (joints) which could have been responsible for the icing.
      3. Ascertain whether or not a fault or a formational contact is located at or near the icing.

4. Prepare a detailed geologic map (1:10,000 or 1:50,000 scale) of the environs of the icing.
5. If the icing originates at a spring, give a complete description of the spring.

B. For an icing formed by the river water, detailed study of that section of the river where the icing is formed should include:

1. Longitudinal profile of the river.
2. Transverse profile of the river.
3. Volume and velocity of the flow.
4. Character of the river bed, shores, valley.
5. Available data on the hydrologic regime of the river.
6. Development of ice along the river and across it.
7. Deformation of the ice, sagging, bulging, etc.
8. Unfrozen windows of water in river ice (Polyn'ya).

13. Other factors which had some effect on the formation of the icing, such as building, ditches, roads, bridges, etc. and natural conditions which are in full or in part responsible for the icing such as distribution of snow-cover, vegetation, climatic factors, soil factors, and others.
14. The study of physico-chemical properties of water and ice should include:

A. Temperature of the icing water at its place of issue. If water flows from several seepages - the temperature of each should be measured. For a river icing, the temperature of the river above and below the icing should be measured. The temperature of the air, in the shade, should be measured at the same time.
B. The discharge of water that feeds the icing. The numerous trickles on the surface of the icing should be converged into a single ditch in which the volume of water is measured. If the study is made during the spring, measurements are made in the morning and at about 2 o'clock in the afternoon in order to determine the increase in the amount of outflow due to melting of the surface ice.
C. Chemical characteristics of water. Samples of water from each seepage are collected for chemical tests. Bottles with samples of water should not be filled to the neck and should be packed for shipment in well insulated containers. Containers with hot water may be packed with the bottle which contains the sample to prevent its freezing. Sample of ice is also sent to the laboratory for chemical analysis.
D. Field studies of ice should cover: color; transparency; amount, character and distribution of air bubbles and of other impurities; the vertical and lateral change in the character, color, transparency, and the structure of ice.

E. If gas bubbles are present at the source of the icing samples of gas should be taken for analysis. The study of gases frequently helps to determine the nature and source of water.

15. Description of the shape and structure of the icing. Whether rounded, elliptical, single layer, of many layers, icicles, cascades, etc. To determine direction of growth of the icing the water should be dyed so that during the subsequent inspection the previous stage of the icing can be recognized by the color of the ice, thus allowing ready determination of the rate and the direction of growth of the icing.
16. The volume and the area of the icing. The map should be supplemented with sketches, photographs and notes on: relief, vegetation, effect on nearby structures, such as roads, bridges, houses, etc. The body of ice should be shown by contours. The map should also show the source, cracks, mounds, areas covered with snow, thickness of snow. Data of subsequent inspections of the icing are plotted on the same map. Icings in forested areas can be recorded by notching the trees and their winter volume can be then determined in the summer.
17. Information should be obtained on the date of the beginning of icing; its constancy in size from year to year, (particularly in relation to dry and wet years), its constancy in position. If at the moment of inspection the icing is dry and no water flows out of it, it is important to ascertain the date when the water ceased to flow, and whether or not the stoppage of flow is temporary or permanent for that season. Reasons for the cessation of flow should be inquired into.
18. Description, measurement, and mapping of frost-mounds and icing-mounds with dates of their origin and changes undergone during the season. Particular attention should be paid to the growth of mounds and the flow of water from their summits or from cracks on their sides. Excavations are recommended to determine the structure of each mound. The system of cracks in a mound should be carefully mapped or recorded with the statement whether or not they extend into the ground adjacent to the mound.
19. Deformation of roads, houses and other structures by the icing and associated mounds should be carefully examined and evaluated.
20. Pertinent factors should be analyzed in connection with the use of the icing water as a source of supply.
21. Measures to prevent the damaging effect of a particular icing should be studied and appropriate recommendations made, particularly if construction of a road is contemplated.
22. Icings which are tentatively chosen for exploitation of water supply require further study during the summer. When the ice is melted and the ground beneath is exposed, a more accurate measurement can be made of its winter volume. The ground itself and its hydrologic aspect are also studied in detail. Final report on the icing should include:

Map of the icing, with profiles.

Geologic map of surrounding area.
Hydrogeologic map of the area.
Table of climatologic data.
Report on chemical composition of water.
Detailed description of the icing and related phenomena.

Survey of Thermokarst Phenomena

Thermokarst phenomena commonly develop on horizontal or gently sloping surfaces, on terraces, or in river valleys.

The survey of the thermokarst should include the following data:

1. Location of the area.
2. Description of major relief features, such as plateaus, valleys, terraces, slopes, etc. with their:
   a. Dimensions.
   b. Steepness of slopes.
   c. Exposure to sun.
   d. Vegetative cover (trees, grass, moss, etc.)
   e. Drainage pattern.
3. Minor features of relief due to thermokarst:
   a. Dimensions.
   b. Presence or absence of asymmetry.
   c. Exposure to sun.
   d. Situation with reference to relief features.
4. Thickness of the thawed ground.
5. Character of the frozen ground.
   a. Whether ground-ice is present.
   b. Temperature of the ground at various levels, to as great a depth as the equipment and time will permit.
6. Study of the geologic structure of the area. In the absence of natural outcrops pits or ditches should be dug, slopes cleared of talus and other excavations made.
7. Study of the ground-water regime.
8. The various data should be mapped, sketched, photographed, and represented in cross-sections and diagrams.

Survey of Permafrost for Buildings

<u>Buildings and Industrial Plants</u>. Surveys of permanently frozen ground preliminary to any engineering construction may be divided into:

1. Reconnaissance survey.
2. Preliminary survey.
3. Final survey.

I. Reconnaissance surveys consist of examining a number of tentatively selected places suitable for construction in the light of local conditions and water supply. This investigation should be started in the spring before the break-up of the ice in rivers, before the melting of the icings, and while the frost mounds are still at their maximum. The following information should be obtained and recorded during a reconnaissance survey:

1. Areas of icing.
2. Source of water that forms icing.
3. Locations of frost mounds.
4. Locations of cave-ins.
   Inhabitants should be asked whether these phenomena are periodic or constant and at what rate they change, etc.
5. Swamps should be located and examined, also the areas of peat and moss. Draining swamps and removing the moss and peat cover will materially affect the regime of the ground. The highest level of water in a swamp should be ascertained.
6. The depth of freezing of rivers and lakes. The extent of freezing of rivers controls the underground percolation of water, which in turn causes deep thawing (lowering of the permafrost table) or provides supply of water for icing.
7. Period of maximum precipitation.
8. The thickness of the cover of snow for the most snowy winters and least snowy winters. (Thin cover of snow, especially during the early part of the winter, indicates possible deep freezing of the ground).
9. The following meteorological data, if available, should be evaluated:
   Air temperatures.
   Precipitation, amount and seasonal distribution.
   The cover of snow, its duration.
   Period of torrential rains.
   Prevailing winds, their velocity.
   The regime of rivers.
   The thermal regime of the ground.
   Texture of the ground.
   Ground-water conditions.
10. Vegetation:
    Pine or fir frequently indicates absence of permafrost or its presence at a considerable depth, whereas larch and brich, particularly if the trees are dwarfed and stunted, more frequently point to the presence of permafrost close to surface.
11. It is valuable at this stage of investigation to prepare a sketch map, even if only with approximate aneroid control, showing relief, locations, and extent of various significant features in the area.

As a result of such a reconnaissance, several apparently equally suitable sites may be selected. These sites are then evaluated with respect to adequacy of water supply, the search for which is carried on during the reconnaissance, or even prior to it. In this connection a tentative plan of pipeline should be considered and its profile sketched.

Attention should also be given to sewage systems and to disposal of industrial wastes, mine dumps, etc.

If the location of a site is already determined by other compelling reasons then there is no need of a reconnaissance survey and one should proceed directly with the final investigation.

Other conditions being equal, preference is given to sites that are:

1. Flat and dry.
2. Not subject to floods.
3. Free of icing, frost-mounds, cave-ins, and swamps.
4. Covered by a relatively thin layer of moss.
5. Distant from mountain slopes (areas near the foot of a mountain have numerous icings, frost-mounds, and ground-ice).
6. Distant from canyons and gullies.

II. Preliminary surveys:

The main purpose of this survey is to select the most satisfactory site of the several that were provisionally chosen during the reconnaissance study.

Preliminary surveys should be carried out during the spring, summer, and early fall and should aim to obtain the following information:

1. A topographic map (with contour interval of not more than 2 meters) of the site and the vicinity to give a picture of the behavior of surface waters and possibility of drainage.
2. Location on the map of proposed pipe-lines leading to the source of water supply with the profile along this pipe-line.
3. Distribution of vegetation (forest, shrubs, grassland, swamp, etc.); the following should be indicated on map:

    a. Dominant kind of trees and their timber quality.
    b. Detailed record of grasses.
    c. Character and thickness of moss and peat.

4. Characteristics of the ground.

    a. Profile or cross-section across site.
    b. Texture of ground.
    c. Amount of moisture (ice).
    d. Water table and general ground-water data.
    e. Hydrostatic pressure of ground water.

f. Thickness (and depth) of frozen ground.
g. Thickness and areal extent of buried ice layers (ground-ice).
h. Underground temperatures.

This study is carried out by means of test drill holes and test pits and by the laboratory study of samples. Samples should be collected at vertical intervals not greater than ½ meter.

Test drill holes should be rather evenly spaced over the area studied and should not be more than 150 meters apart. Additional bore holes should be placed at the breaks of slopes, along gulleys, canyons, and near lakes and rivers.

Bore holes should be at least 5 meters deep. If within this distance the drill strikes bedrock it will not be necessary to continue the hole. However, as an occasional boulder may be mistaken for bedrock test pits should be dug at frequent intervals to control and to amplify the data obtained from the drill holes and to provide better samples for more detailed analyses. The number of test pits should be at least 20% of test drills. Drill holes are also utilized for measuring ground temperatures.

In this preliminary investigation it will suffice to read temperatures at a given time in some 15 or 20 holes situated in varied environments. Temperatures are read at depths of 1.6, 2.5, 10 and 20 meters. Temperatures at depths of 10 and 20 meters are measured in the holes that are drilled for that purpose. If these temperatures show little or no variation from place to place it is then not necessary to drill more than 3 or 4 twenty-meter holes. If, on the other hand, temperatures vary by more than one degree and if the frozen ground contains taliks, more detailed investigation is necessary. Temperatures should be measured once every 10 days and those at the depth of 10 and 20 meters, once a month.

If there is a building or any other structure near the site the regime of this structure should be carefully investigated. This should include ground conditions as well as structural design, particularly that of the foundation. It is also necessary to draw a contour map of the surface representing the bottom limit of the thawed ground adjoining the building. It should be based on soil auger tests spaced at 1, 3, 5, and 10 meters from the building, on both the north and south sides or, preferably, on all four sides. These observations should be made at least twice a month.

If possible the depth of the thawed ground should also be measured beneath the building and the material and construction of the floor should be noted.

The most dependable and stable ground for any construction is solid bedrock at or near the surface. Such rock should be free from fissures (joints), particularly those that carry ground water or are filled with ice.

If the bedrock surface lies within or below the active layer its value as a stable base for a foundation is greatly decreased. The winter swelling of the active layer together with the difficulty of anchoring a foundation, especially if the bedrock is fissured, will result in damage to the structure.

Sites that are free of frozen ground - islands of unfrozen ground that are occasionally found in the permafrost province, are not always the most satisfactory. Unfrozen swampy and unstable ground is as a rule less secure and less economical than an area with firmly frozen ground. Equally unsatisfactory are areas free of permafrost but containing strongly swelling ground, as where the active layer (seasonal freezing) is very thick.

In areas of permafrost the following conditions are regarded as necessary to insure stability of an ordinary building:

1. A considerable thickness of permafrost - not less than 25 meters.
2. A more or less constant sub-zero temperature from a depth of 10 meters downward (temperature should not be higher than -1.0° C.)
3. Permafrost with sufficient moisture (ice), close to, but not exceeding saturation.
4. Homogeneous and evenly-thawing active layer where this layer does not exceed 2 or 3 meters.
5. An active layer that does not swell.
6. Active layer preferably composed of well-drained gravelly material rather than of clay or silt.

Areas with layered permafrost are unsuitable and should be avoided. However, if no other kind of ground is available a thorough investigation of ground conditions should be made before a structure is erected. Equally unsuitable are places with extensive icing, frost-mounds, and where permafrost contains layers of ice not far beneath the floor of a projected excavation.

III. Final investigation:

Final investigation should begin after the completion of preliminary studies and should be carried out in the fall and through the winter. Under favorable conditions this investigation may be completed by August or September of the following year.

By this time the survey of the water supply should be completed.

It is desirable that the person who is to be in charge of the completed project and who is, therefore, well informed on the function and behavior of the structure, should participate and cooperate in this final investigation.

The purpose of the final investigation is to supplement the information already obtained during the preliminary survey so as to furnish all the necessary data for planning and computation of specifications of the projected structure. Thus the final survey calls for:

1. More detailed topographic map of a more restricted area - the actual site - with the contour interval of about 2 feet (half a meter).

2. More detailed information on the ground-water regime. Additional test holes may be necessary. These holes should be drilled between the middle of August and October - the time of maximum seasonal thawing. Particular attention should be paid to holes with artesian flow.

Pressure - surface (isopiestic) and ground-water table maps should be made for at least two periods (early September and early November). It may be very difficult to make these maps as there may be one hole with strong artesian flow yet another hole only 2 or 3 meters away may be absolutely dry. Such occurrences are, in all probability, due to the existence of an irregular network of subsurface channels forming a confined aquifer whose water is under considerable hydrostatic pressure. As yet little is known about this aspect of ground-water in the permafrost province.

Particular attention should be given to the presence, amount, and the direction of ground-water movement.

Moisture content of the ground (in the form of water or ice) should be determined for the active layer and for the upper part of the permafrost. These data are necessary for computing the adfreezing force. The amount of ice in the permafrost, expressed in percentage will be used as a basis for various computations including the amount of expected settling of the ground when thawed and the amount of stored cold.

3. More detailed record of thermal regime of the ground. This should include:

   a. Amplitude of temperature variations in the active layer and in the upper part of the permafrost.

   b. The time lag of temperatures with depth.

   c. Duration of the frozen condition in the active layer at different levels.

   d. Distance to the layer of zero annual amplitude and the magnitude of the geothermal interval or gradient.

4. More accurate data on permafrost table to be obtained by the use of soil auger. A contour map of the permafrost table is prepared and cross sections drawn showing the distribution of the kinds of ground and the position of the permafrost table.

5. Detailed analysis of the nature and amount of swelling of the active layer [6]

Meanwhile, as outlined by Yanovsky, the following studies and tests should be made in the laboratory:

A. Physical properties of the ground:
   1. Mechanical composition.
   2. Specific gravity of the ground.
   3. Maximum moisture content when compacted.
   4. Angle of repose.
   5. Bearing strength of the ground.
   6. Coefficient of internal friction.
   7. Coefficient of filtration (permeability).
   8. Compressive strength in relation to moisture content.
   9. Properties of the ground in a frozen state:
      a. Crushing strength.
      b. Coefficient of swelling (on freezing).
      c. Adfreezing strength.

In October, after all these observations have been completed, the building site is stripped of vegetation and sod, and is appropriately graded. During the winter the snow should be kept off the surface and in August or September of the following year all observations on the depth of permafrost table should be repeated and additional test pits examined. The observed change in the level of the permafrost table should be accurately recorded.

Survey of Road Routes

The frozen ground phenomenon that cause the greatest difficulties in road construction and maintenance are the swelling of ground, icings or "ice fields", frost boils, cave-ins, and landslides. Icings are directly connected with streams, rivers, and freely percolating ground-water whereas frost boils and cave-ins are caused by the melting of ground-ice.

---

6 Differential swelling of the active layer, may be measured with a very simple device, called a swellometer, a diagram of which is here reproduced from Bykov and Kapterev. (Fig. 86). The swellometer consists of eight rods each of which is firmly fastened to a hollow prismatic base and a ninth rod set in a solid base. The rods are greased and assembled in such a way that they may slide freely and spread upward. Care should be taken not to apply any grease on the outside surface of the basal parts. These should be left rough or may be notched or barbed to obtain a firm bond with the surrounding frozen ground that is being tested. Reading the swellometer should be done with a telescopic level and referred to some fixed point or bench mark that will not, in any way be affected by the swelling of ground. (Fig. 87).

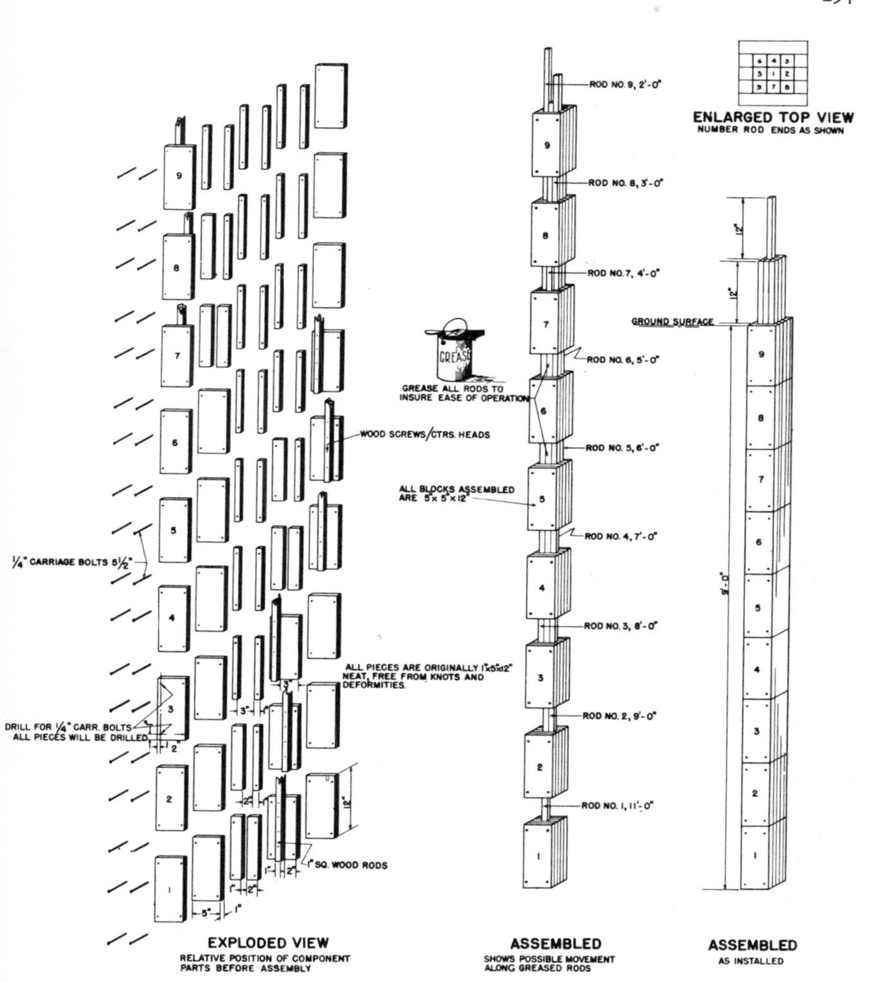

DESIGN OF SWELLOMETER

(AFTER BYKOV AND KAPTEREV)

FIG. 86

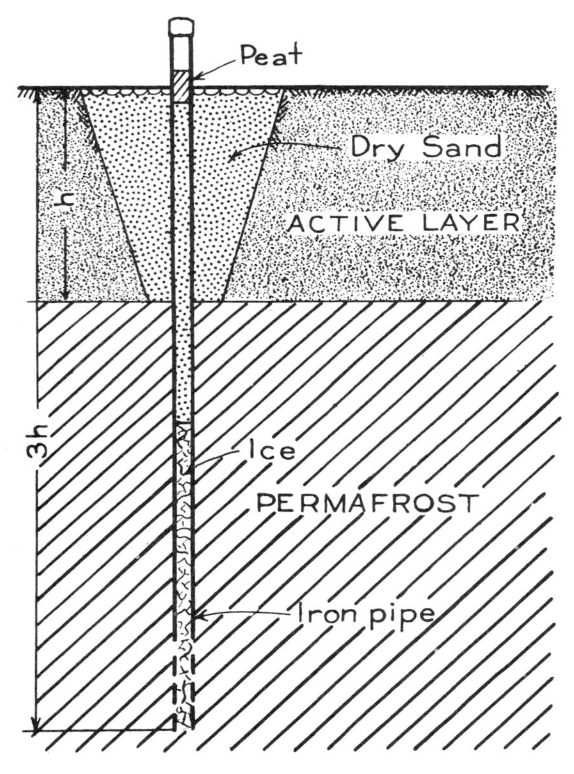

SETTING OF BENCHMARK (Bykov and Kapterov)

Metal pipe perforated at base is imbedded in ground to a depth three times(3h) the thickness of active layer (h).

FIG. 87

The first steps in a survey of a road route should, therefore, consist of the following investigation:

1. Determine whether ground-water is present or absent. If present:
   a. Amount of ground-water.
   b. Direction of flow or percolation.
   c. Thickness and other geologic aspects of water-bearing layers.
   d. Possibility of drainage or diversion of ground water.
2. Test drill to determine whether underground ice is present or absent: For fills it is usually sufficient to test the ground to a depth of 1 to 1.5 meters below the active layer. However, in swamps and in peat bogs where, as a result of drainage and clearing of the surface, the active layer may subsequently increase in thickness, drilling should be carried to a depth of at least 4 meters. For road cuts (and other excavations) the ground should be similarly tested to a depth of 4 meters below the lowest level of the surface of the projected excavation. These test holes should be about 50 meters apart along the projected route, except where bedrock is encountered.

The presence of ice inclusions necessitates a more detailed examination of the ground. A projected roadbed should be tested through its entire width and about 10 meters beyond on each side, with the test holes spaced not more than 15 meters apart. If ice inclusions prove to be of considerable thickness and areal extent, a survey of the ground should be extended over a sufficiently wide area to determine the possibility of by-passing the bad stretch.

Wherever possible the ice layers and lenses should be contoured and plotted on the map with their thicknesses.

Large inclusions of ice are usually easily detected but those 0.5 to 2 meters thick and of sporadic distribution may easily be missed. In construction of a fill these smaller masses of ice that are accidentally overlooked during the preliminary survey are negligible. Only in exceptional cases, with the unusual lowering of the permafrost table, do they effect the roadbed. However, in road cuts and other excavations these undetected ice layers are likely to result in serious cave-ins. During the first year of operation of the road it is essential that serious attention should be given to a timely discovery of such defects and to their immediate repair.

In laying out a road traversing areas of silts or sandy-clayey ground, fills should be preferred to cuts, which should be avoided altogether if possible. But at the same time fills should not be made excessively high.

No definite information is available on the relative advantages of utilizing either the north or the south-facing slopes of the country. Some investigators, however, have observed that in an east-west valley the south-facing side tends to have a more pronounced break of slope than the north-facing side. This difference is ascribed to a deeper penetration of summer thawing and consequent flowage of ground along the south-facing slope.

In constructing fills, cuts, and in making other excavations the texture and moisture content of the ground is studied. These properties determine the angle of repose of cuts and fills and also the suitability of certain ground for road metal.

Such projects should take into consideration the supply of suitable road metal because local ground in many areas consists of as much as 95% of silty material, which is unsuitable for fill.

Survey of Bridge Sites

Construction of bridges over gulleys, small streams, and rivers calls for a detailed preliminary examination of frozen ground conditions. Permafrost table near a flowing stream may be at a considerable depth and may not enter into calculations at all. Critical factors here are the depth of seasonal freezing - (thickness of active layer) and the swelling of ground.

Preliminary survey of a bridge site should furnish:

1. A detailed cross-section of a gulley or river and the approaches showing:
   a. River bed.
   b. Underflow conduit.
   c. Active layer beneath the stream and on the approaches.
   d. Permafrost table.
   e. Geologic structure of the ground.
   f. Composition and texture of the ground.

This cross-section should extend to a depth of at least 4 or 5 meters below the base of the foundation of a projected structure unless solid bedrock is encountered before this depth is reached.

In the presence of bedrock, in addition to the necessary number of bore holes to determine the character of the overlying ground, one or two pits are dug so that the character and thickness of bedrock may be examined in greater detail. Such pits should penetrate bedrock to a depth of at least 1.5 meters.

Depending upon the size of the projected structure it may become necessary to construct one, two, or more cross-sections at right angles to the one described above.

Large structures that require sinking of piers to a considerable depth call for a more thorough and more extensive preliminary investigation.

Survey of Airfield Sites

The following procedure for the survey of airfield sites, preliminary to the final selection of a site, is to supplement the general instructions given in the "Aviation Engineers", pages 59-62.

STAGE I

a. Preliminary study of available (published) sources on

(1) Terrain (geology) and soils, occurrence and distribution of permafrost.
(2) Climate.
(3) Vegetation.
(4) Hydrology - seasonal changes of streams.
(5) Accessibility.

b. Flight reconnaissance, study of terrain from the air by flying officer, geologist or soils engineer, and a construction engineer.
c. Aerial photography.

(1) Vertical strip from 10,000 feet above the ground using 12 inch cone giving a scale of approximately 1/10,000.
(2) Selected oblique views of suitable areas.
(3) Vertical and oblique views of possible access routes.

STAGE II

d. Preliminary ground surveys of suitable sites (with hand tools and probing rods) by the same personnel who participated in the flight reconnaissance, item b. in Stage I. Due to the prevailingly swampy nature of the terrain and the presence of snow cover in the winter, these preliminary ground surveys should be done in summer and early fall.

(1) Examination of ground exposures by soils engineer or a geologist.
(2) Preparation of a cover map, showing vegetation and type of top soil, on a vertical mozaic, by soils engineer or a geologist. Services of a botanist are desirable.
(3) Study of possible sources of water supply.
(4) Gathering of other pertinent data.

(a) Prevailing winds.
(b) Obtain ground evidence on floods (driftwood, bark trees, etc.). Compare with the data obtained in Stage I, a. (4).
(c) Record thickness of seasonal frost and the amount of thaw.
(d) Other observations on permafrost phenomena.

(5) Availability of construction materials (gravel, sand, timber, etc.).

STAGE III

e. Preliminary ground survey with drilling equipment. A limited number of test holes to be located at critical places to determine whether or not dangerous or unsuitable ground is present.

(1) Active Layer

(a) Thickness.
(b) Variations in thickness from place to place.
(c) Type of soil determined from visual inspection.
(d) Sampling of soil for laboratory tests, including field moisture, field density, texture, and other standard tests.

(2) Permafrost

(a) Depth to permafrost table.
(b) Thickness of permafrost, if not excessively thick (not over 50 to 60 ft.), can be pierced during the preliminary drill tests.
(c) Type of permafrost - whether continuous or discontinuous laterally and whether continuous vertically or layered.
(d) Composition and vertical profile of permafrost based on visual inspection.
(e) Surface and underground extent of unfrozen ground (taliks).
(f) Ground-ice, ice lenses and wedges, their thickness and extent.
(g) Observations on ground-water table.
(h) Presence of water-bearing layers or channels of unfrozen ground (taliks).
(i) Sampling of permafrost for laboratory soil tests.

(3) Relief features connected with frost phenomena which are indicative of potentially destructive character of the ground.

(a) Frost-mounds.
(b) Cave-in lakes.
(c) Cave-in funnels (thermokarst).
(d) Icings, or sites where icings had been observed during the winter.
(e) Polygonal soil markings.
(f) Slump and landslides.
(g) Creep and mudflow-like behaviour of ground, solifluction.

f. Permafrost survey with geophysical methods.
g. Preparation of permafrost map embodying all the above information
h. Preparation of topographic map with contour interval of 5 or 10 feet.
i. Comprehensive survey for runway and other installations.

Survey of Water Resources

Ground Water. Investigation of the ground-water supply in the permafrost province may be divided into:

1. Reconnaissance survey.
2. General survey.
3. Special survey.
4. Stationary survey.

An important step in the preliminary work in all ground-water surveys consists in obtaining and compiling pertinent meteorological data for the area to be surveyed. These data may be summarized in tabular form as shown on page 160.

The following field equipment is essential for reconnaissance and general surveys of water resources:

1. Portable field laboratory for chemical analysis of water.
2. Equipment for the determination pH.
3. Thermometers.
4. Stop watch.
5. Current meter.
6. Equipment to test the permeability of ground.

Field investigation of ground-water resources should be carried out with the geologic map as source of basic data. In examining rock formations and separate outcrops particular attention should be paid to the following:

1. The amount of jointing (fissuring) in the rock, whether fissures (joints) are open or sealed. Material that fills the fissures should be determined, whether it is mud, ice, or calcite, etc. The size of fissures, their orientation, relation to bedding, and to relief should be observed and recorded. Fissures produced by the frost action should also be noted. Frost fissures may extend one meter below the surface of the ground and may be traced along the surface for several tens of meters. In the summer these fissures are usually sealed and can easily be overlooked. The best time to observe these fissures is in the autumn (September to November) and early in the spring (February to April). The presence of polygonal frost-cracks has great hydrologic significance as they permit the rain and surface water to seep down into the water-bearing horizons above the permafrost. They also permit lateral circulation of water which, upon freezing, forms ice-veins or ice-wedges.
2. Porosity of the ground and incidence of caves. Large caverns should be thoroughly explored - their size, orientation, relation to relief, to fissures (jointing), to the composition of the rock and their position with reference to their present and past

base-levels of erosion. In large caverns with flowing water the temperature of the air and of the wall rock should be measured. The presence of ice masses should be noted and their origin should be ascertained.

3. Constancy of stratigraphic horizons includes the study of lateral changes in the character of the rock or ground particularly with regard to porosity or permeability.
4. The presence of various spring deposits (iron, lime, ice, etc.).
5. Bleaching, traces of encrustation, and other indications of the presence of moisture.
6. Products of weathering, their character, extent, and the composition of the residual material.
7. Quaternary deposits and Recent alluvium. These should be thoroughly investigated, particularly in relation to the extent of permafrost.

The following observations should be made with regard to the plant and animal life:

1. Trees growing on a slope may have curved trunks indicating the creep (or solifluction) of the ground along the permafrost table. Orientation and areal distribution of curved trunks should be noted and the age of straight and curved trees should be determined.
2. "Drunken forest" (Fig. 31) - a group of irregularly inclined trees - normally indicates the presence of frost-mounds or strongly swelling ground.
3. Willow groves point to the presence of ground-water which freezes only for a short time.
4. Trees on pingo (large frost-mound) may enable one to determine how old the mound is.

   In the permafrost province the depth of tree roots approximately corresponds to the thickness of the active layer making it possible to get some idea of the thickness of the water-bearing zone above the permafrost.
5. Peat and moss usually indicate a relatively thin water-bearing zone above permafrost.
6. Pine and fir commonly grow where permafrost is either absent or lies at a considerable depth.
7. Larch and birch, especially dwarfed and stunted individuals, point to the presence of permafrost close to the surface.
8. In steppe areas various other plants may be used as indicators of the presence of ground-water close to the surface.

Holes of certain burrowing animals should be examined carefully as they usually occur where the water above the permafrost begins at some depth below the surface of the ground. The material that is thrown out of the hole by the animal may furnish information as to the composition of the active layer.

The study of relief of the area has a direct bearing on the solution of certain ground-water problems. The following features of relief should be examined:

1. The character of terraces, slopes, and valley floors.
2. Positive elements of relief that are produced by frost action (frost-mounds, pingos, "pseudo-terraces", etc.) should be examined and their areal distribution shown on a map.
3. Cave-in lakes, funnels, sink-holes, etc. due to permafrost phenomena should be studied and mapped.
4. Landslides, slumps, and solifluction should be similarly studied, also karst phenomena and frost cracks in the soil.

Other features to be studied in connection with the survey of water resources of an area include:

A. Icings.
B. Springs and polyn'yas (singular - polyn'ya - a hole or a window in river ice, which remains unfrozen during all or part of the winter due to a local inflow of warmer water, as from a subaqueous spring or from a warm tributary).
C. Wells and drill-holes.
D. Test pits, ditches, and other excavations.
E. Water bodies such as rivers, lakes, swamps, etc.

The amount of detail with which each of the above items should be studied will be determined by the purpose of the survey or by the nature of the assignment.

Springs. Each spring should be numbered, mapped and described. The notes should contain:

1. The name of the spring.
2. Location, particularly in relation to the surrounding relief and its exposure to sunlight.
3. Altitude and also the height above the nearest large body of water (river or lake).
4. Geological surroundings:

   A. Character and age of rocks which serve as the source of the water.
   B. The outlet. Whether in an unconsolidated alluvium, coarse talus debris, etc.; whether along a bedding plane, in a joint (fissure), a fault plane, a cavern, a contact between two different formations, or in a ground-ice. If in a pingo, whether at the summit or along its slope, in the crater or a side fissure.

Presence or absence of swamps in the vicinity of the spring should be noted. It should be also stated whether the water comes out to the surface in a vigorous flow, or seeps out of the rock.

In the permafrost area springs occasionally issue at the summits of hills, through cracks in the soil, or at the junction of two valleys. Frost-mounds, "drunken forests", broken-up soil, and splintered tree stumps are commonly present in the vicinity of a spring.

A spring usually issues where the water table comes to the surface of a sloping ground or it may be fed by waters from a confined aquifer. The first problem, therefore, is to determine the origin of the spring. The solution of this problem will furnish the data which is necessary to ascertain the extent of water-bearing layers and the possible location of other springs.

5. In the permafrost province it is very important to determine whether or not the outlet of a spring is always in the same position. Springs are known to have shifted their outlets tens and even hundreds of meters from year to year or during a single season. The shifting of the spring should be described, giving the direction and the time of the shift and stating whether the shift is progressively in one direction and regular or sporadic.
6. The type of the spring: gravitational seepages are usually formed by water above the permafrost, whereas the upwelling springs are fed from below the permafrost. It should be borne in mind, however, that the water below the permafrost may rise to the surface at the foot of a mountain and percolate through the surficial ground down the slope to the river bank where it may emerge as a gravitational spring. During the winter such a spring is likely to shift its position, up the slope towards the foot of the mountain, nearer the place of its original source.
7. The yield of a spring. The yield of a spring, the time at which it was measured and the method employed should be described. Several measurements of the yield should be made in order to determine the constancy of the flow. The most important time to measure the flow of a spring is in the spring (February and March), when active layer is frozen to the maximum depth. Other measurements should be made in September to October at the time of the maximum thawing of the active layer. Some springs cease completely at this time whereas others reach their maximum yield. In the spring (February - March) while measuring the yield of a spring observations on icings (nalyed) should be made at the same time. The amount of ice formed in an icing as well as the date of its inception should be correlated with the flow of the spring. A comparative study of several springs with their icings will furnish the necessary data for the selection of the most suitable source of water supply.

In the permafrost province the flow of springs is usually quite variable from season to season so that in order to obtain a complete record of a particular spring repeated measurements of the flow should be made. The critical period of a spring is from January to May. During this time some springs diminish in their flow and occasionally cease to flow altogether whereas other springs, on the contrary, increase in their flow. An interesting phenomenon, called "the opening of a spring" can be observed between March and June at some of these springs which freeze completely during the winter. A crack or a funnel appears at or near the frozen spring and a vigorous flow of water suddenly starts from this opening. The formation of a crack is frequently accompanied with a sound which resembles a gun shot. The flow of such a spring, immediately after its opening, is usually considerably above normal, gradually diminishing as time goes on.

8. Physico-chemical properties of water. The temperature of water should be measured at the outlet of the spring. The temperature of the air and the barometric pressure should be recorded at the same time. Temperatures should be measured in the shade.

Simple chemical tests of water are made with what facilities are available in the field and samples are sent to the laboratory for a more complete analysis. Similarly the chemical precipitates (solids) and gases are examined in the field and samples forwarded to the laboratory.

Owing to the winter freezing of the surface ground the amount of gas in a spring may vary from season to season. During the winter freeze the gas is unable to escape through the surrounding ground and therefore concentrates in the partly frozen constricted vent of a spring. The laboratory tests of gases which are dissolved in water may throw some light on the origin or the source of the water.

The surface water in rivers and swamps should be examined for gas bubbles as their existence may indicate the presence of a bottom (subaqueous) seepage of ground water. The presence of such a spring in a river can be detected in a number of different ways. A sudden increase in the flow of a stream in a part of a channel where there are no tributaries is a sure sign. This can be confirmed by examining the differences in temperature, gas content, and chemical composition of water above and below the spring. In smaller streams which in the winter usually freeze through to the bottom a subaqueous spring will work its way to the surface of the river ice and will flow out and freeze in successive sheets producing an icing. Large springs may reveal their presence by delaying or completely preventing the winter freezing of the river at the place of their issue. The resultant window-like hole in the river ice is called polyn'ya by the Russians.

Springs may also appear in a dry river bed and, after flowing on the surface a short distance, may disappear again into the ground. Such springs indicate the presence of water above the permafrost and should be recorded and mapped.

Radioactive tests may serve a useful purpose as the spring which are fed by the water from below the permafrost are usuall relatively high in radioactive elements.

9. Hydrotechnical data should include such pertinent information a whether the spring is developed or not. If developed, a detailed description of construction, upkeep, and accessibility should be given.
10. Sanitary description of a spring should give the distances to inhabited houses, cesspools, and other such places that are likely to cause contamination of water. It should be stated whether or not the spring is likely to be contaminated by the flow from the street gutters, stables, etc., whether cattle are watered at the spring, and whether the laundrying is done at the spring. If available, the bacteriological analysis of water should also be given.
11. Exploitation of springs. The report should state whether or not the spring is being used by the local population, and if so, whether or not it furnishes a sufficient amount of water throughout the year. If water is used only during a part of the year, the reason should be explained. The report should also include the names of the owner, the person in charge, the builder and give the condition of weather on the day of inspection of the spring and on the immediately preceding days.

Examination of Water Wells. During the survey of water resources all water wells, whether dug or drilled, are studied, plotted on the map and recorded.

In addition to the observations outlined in the study of springs, the examination of a well should also include:

1. Size of the well.
2. Height of the casing above the ground.
3. Depth of the well to the bottom and to the water level.
4. Geologic cross-section of the well.
5. Description of aquifers.
6. Whether or not the well empties easily.
7. How quickly the water level is restored during the pumping.
8. How constant is the regime of the well.
9. Factors which control the oscillation of water level.

Just as in the case of a spring, the report on the well should contain the hydrotechnical and sanitary data. In addition all details should be obtained on the equipment that is used to lift the water (pumps, buckets, etc.).

The most valuable information that is furnished by the study of a well is the level of the water in the well and the depth of the well. These measurements furnish the data on the depth of the water-bearing layers above and below the permafrost and, if made repeatedly, may give some data on the fluctuation of these aquifers. The evaluation of these measurements should be made with some caution, as the standing level of water in a well may not necessarily reflect the true level of ground water. Occasionally the bottom or walls of the well may become frozen and the water may thus become isolated from ground-water source. In drilled wells water frequently freezes in that part of the well which is surrounded by the permafrost and the water above this ice plug becomes separated from the aquifer. The measurements of water level in such a well, obviously, have no value. To make sure that the well water is connected with the aquifer it is necessary either to pump out the water from the well or to pump in additional water and to observe the water level. The presence of an ice plug may also be established by measuring the depth of the well if the depth to which the well was originally sunk is known.

It should be borne in mind that the winter fluctuation of the water level above the permafrost is caused by the freezing of surficial ground. The rise of water level in a well or its overflowing during the winter therefore is not a sign of an increase in the amount of ground-water supply but, on the contrary, is an indication that the supply of ground-water is reduced due to the freezing of a part of the aquifer. The rise of the water level in the well is due to the hydrostatic pressure which develops in the aquifer when parts of it become frozen and the gravitational percolation of water is then driven with force as in a confined aquifer.

Wells that are fed by an uninterrupted inflow of ground water afford the best means of studying the fluctuations of the ground-water level.

In some shallow wells, dug through a clayey ground, the casing is likely to be affected by frost heaving. The casing and the immediately surrounding ground do not remain stationary, making it difficult to record the water level in the well. In such cases a careful observation should be made of the movement of the casing as well as of the water level in the well and readings should be recorded with reference to some benchmark that is not subject to frost-heaving.

The most satisfactory results in a ground-water survey are obtained through the study of drill-holes or test pits made especially for this purpose.

Description of a well should contain observations on the deformation of well casings, pump houses, water mains, etc. Permafrost phenomena in the vicinity of the well such as frost-mounds, icings, and so on, should also be described.

<u>Examination of Drill-holes</u>. Examination of a drill-hole should furnish the following information:

1. Thickness and temperature of the ground above the permafrost.
2. Thickness and temperature of the permafrost.
3. The number of penetrated aquifers.
4. Thickness and rock character of each aquifer.
5. The stationary level of water.
6. Geologic cross-section of the area showing the positions of aquifers.

7. The yield of the well as determined by test pumping, with the data on the lowering of the level of water during the pumping.
8. Analysis of the water, temperature, etc.

Survey of Surface Water Supply. Rivers, creeks, and lakes should be measured in length, width, and depth. All the changes in the river bed and in its banks should be described proceeding down the stream.

The mean level of the river and its flood height should be ascertained.

Discharge measurements should be made at frequent intervals in order to determine the amount of water lost by filtering through the ground or the amount added to the stream by the ground-water seepages.

In the study of surface water bodies the chief points of study are:

1. The regime of the water body (volumetric and thermal).
2. The source of its water.
3. The relation to ground water and to surface waters.
4. Whether or not the river or lake freezes through to the bottom and whether or not bottom-ice is formed.
5. Thickness of ice, the depth of the unfrozen part of the river or lake.
6. Velocity and volume of water beneath the ice.
7. Time of freezing and of breaking up of ice.
8. The manner of the break-up of ice and its duration.
9. Whether or not polyn'yas (unfrozen windows) are present during a part or all of the winter. Polyn'yas should be accurately located on the map for they may indicate the presence of a subaqueous spring or an accelerated flow of water in that part of the river.
10. Sections of rivers with icings and icing-mounds should be examined and each icing as well as the entire area of icings examined measured and plotted on the map.
11. Whether or not ground-ice is present in the banks of the river, and if present, how much water will its melting contribute to the flow of the stream.

Along with the study of rivers and lakes the nearby artificial reservoirs should also be examined and described, giving their location, dimensions, and the type of construction.

In conclusion, the survey should cover all the pertinent data regarding the utilization of the supply, as for example:

1. Conditions of supply and utilization.
2. Conditions of irrigation and drainage.
3. Measures to prevent the damage by icings.
4. Construction of pipe-lines.
5. Development of springs, etc.

The following illustrations should accompany a report on water resources of an area:

1. Geomorphic or relief map of the area.
2. Geologic map.
3. Map of springs, wells, drill-holes, icings, frost-mounds, "drunken forest", ground-ice, etc.
4. Map of waters above the permafrost. This map is usually prepared on a topographic base. It should show areas where the water above the permafrost is present, and where it is absent, giving the time (month) for which the map is compiled. Areas where the water above the permafrost does not freeze through during the winter should be shown. The map may also show sources of such water and its distribution in relation to unconsolidated deposits.
5. Map of the permafrost may show the permafrost table and its case by contour lines of different color or pattern.
6. Map of water below the permafrost usually prepared on a topographic or a geologic base, should show depth of this water below the surface of the ground and the types of the water, whether in a bedded aquifer, solution channels, joints (fissures), etc.
7. Map showing the distribution and thickness of ground-ice and of taliks (unfrozen and usually water-bearing layers).
8. Map of mineral deposits and construction material.
9. Cross-sections, profiles, diagrams and graphs.

Outline of Pipe-line Survey. Location of water mains should be decided upon only after the following investigations of ground along the projected lines have been completed:

1. Hydrologic condition of the ground.
2. Vegetative cover of the ground.
3. Thermal condition of the ground.

   Temperatures should be measured in each different type of surroundings, at least, at 3 different levels (0.8, 1.6, and 3.2 meters). Temperatures are read once every 5 days throughout the year. If a survey throughout the year is not feasible the measurement of temperature can be limited to the period of time when these temperatures at 2-3 meters below the surface are at their minimum - usually from March to May.

   When water mains are planned along projected roads or streets, the ground around the thermometer should be kept clear of snow and vegetation. Data on the composition of ground and condition of ground surface should be recorded along with the temperatures.

Feasibility of draining the surface of ground along the projected water mains should always be considered in the course of a survey.

Test pits should be dug at critical places along the pipe-line to determine the composition and structure of the ground to a depth of about 3 meters. Results may be recorded in a form of a detailed section or a table showing the thicknesses of various layers, especially the thicknesses of frozen ground exposed in a pit. Places where pipe-lines will cross rivers, gullies, roads or streets should be surveyed in a more detailed manner and should be shown on accurately drawn maps and cross-sections.

Should it become unavoidable to run a pipe-line across a swamp attention should be given a possible source of material for a fill.

Final report on pipe-lines should include:

1. Detailed topographic map of a strip along the projected pipe-line.
2. Profile of the lines with an indication of places that are to be excavated or filled.
3. Transverse profiles of rivers, gullies, etc. where these will be crossed by the projected pipe-lines.

All profiles should show water table, permafrost table, thickness of active layer, depth of swamps, and other pertinent data.

GLOSSARY

Terms pertaining to the frozen ground phenomena
(Terms in lower case are provisionally regarded as unnecessary synonyms)

ACICULAR ICE(Fibrous ice, satin ice) - formed at the bottom of ice (near the contact with water); consists of numerous long crystals and hollow tubes of variable form having layered arrangement and containing bubbles of air.

ACTIVE LAYER (Annually thawed layer) - layer of ground above the permafrost which thaws in the summer and freezes again in the winter. (Equivalent to seasonally frozen ground.)

ACTIVE METHOD (of construction) - method of construction in which permanently frozen ground is thawed and kept unfrozen at and near the structure.

ACTIVE PERMAFROST (Active permanently frozen ground) - permafrost which, after having been thawed due to natural or artificial causes, is able to return to permafrost under the present climate.

ADFREEZING - the process by which two objects adhere to one another owing to the binding action of ice as result of freezing of water.

ADFREEZING STRENGTH - resistance to the force that is required to pull apart two objects which adhere to one another as a result of the binding action of freezing. (In Russian reports this term is frequently used to mean TANGENTIAL ADFREEZING STRENGTH).

agdlissartoq - Eskimo name for a frost-mound, lit., "the one that is growing", Pingorssarajuk.

AGGRADATION of PERMAFROST - growth of permafrost under the present climate due to natural or artificial causes. Opposite to degradation.

ANCHOR ICE (Bottom ice) - ice formed on the bottoms of rivers and lakes.

annually thawed layer (Active layer) - a layer in the ground above the permanently frozen ground which is alternately frozen and thawed each year.

APPARENT SPECIFIC GRAVITY - volumetric weight.

AQUIFER - a geologic formation or structure that transmits water in sufficient quantity to supply pumping wells or springs.

aufeis - German term for ICING - (Flood-ice, "Glaciers", "Glaciering"?).

BERM - a bench or a horizontal ledge partway up a slope.

bodeneis - German for ground-ice.

boolgoonyakh - frost mount (pingo, "hydrolaccolith") - a mound, usually of a considerable size and of many years duration - not a seasonal frost mound.

BOTTOM ICE (Anchor ice) - ice formed during the winter on the bottoms or rivers and lakes.

CAPILLARITY - the property of tubes with hairlike openings, when immersed in a fluid, to raise (or depress) the fluid in the tubes above (or below) the surface of the fluid in which they are immersed.

CAPILLARY FRINGE - the zone immediately above the water table in which water is held above the ground-water level by capillarity.

CAPILLARY INTERSTICES - openings small enought to produce appreciable capillary rise.

CAPILLARY WATER - water that is retained in the capillary interstices of the ground and is capable of movement through capillary action. It may remain unfrozen at temperatures between -4° C. to -78° C.

CAVE-IN LAKE (Kettle lake, kettle-hole lake) - a lake formed in a caved-in depression produced by the thawing of ground-ice (ice lens or ice pipe).

CLOSED SYSTEM - a condition of freezing of the ground when no additional supply of ground-water is available.

COMBINED WATER - water of solid solution and water of hydration which does not freeze even at the temperature of -78° C.

COMPACT CRYSTALLINE ICE - ice formed by quiet freezing of water in large basins.

CONFINED GROUND WATER - a body of ground-water overlain by material sufficiently impervious to sever free hydraulic connection with overlying ground-water except at the intake. Confined water moves in conduits under the pressure due to difference in head between intake and discharge areas of the confined water body.

congelating stress - adfreezing strength.

constant soil congelation - permafrost.

constantly frozen ground - permafrost.

CRITICAL MOISTURE CONTENT - maximum amount of interstitial water which, when converted into ice, will fill all the available pore space of the ground.

crystocrene (Icing) - surface masses of ice formed each winter by the overflow of springs.

crystosphene (Ground-ice) - mass or sheet of ice developed by a wedging growth between beds of other material.

DEEP-SEATED SWELLING - swelling of ground caused by the freezing of freely percolating ground water.

DEGRADATION OF PERMAFROST - disappearance of the permafrost due to natural or artificial causes.

deposited ice - bottom-ice.

depth of seasonal change - level of zero annual amplitude.

DILATION, WATER OF - water in excess of water of saturation held by the ground in an inflated state (water of supersaturation).

DITCH WATER - is water of air temperature from streams or reservoirs used in the gold mining in Alaska to thaw the frozen ground during the warm season of the year.

DRY FROZEN GROUND - ground with temperature below 0° C. but containing no ice.

DRY PERMAFROST (Dry permanently frozen ground.) - permanently frozen ground with temperature below 0° C. but containing no ice.

DUFF - the vegetable matter which covers the ground in the forest, as leaves, twigs, dead logs, etc.

earth-mound - frost-mound.

eisboden - German for frozen ground.

eis als felsart - German for ground-ice.

eis im boden - German for ground-ice.

eruption of soil - ? frost-mound.

eternal frigidness - permafrost.

eternally frozen ground - permafrost.

ever frozen soil, subsoil, or ground - permafrost.

FIBROUS ICE - acicular ice.

FINE AGGREGATE ICE - ice formed by freezing of stirred water.

FIRN ICE - formed by freezing of snow into separate spherical granules of dull appearance

FISSURE-POLYGONS (Mud-polygon) - gently convex polygonal areas of ground separated from each other by grooves or fissures. Includes TUNDRA POLYGONS and MUD-FLAT-POLYGONS.

FIXED GROUND WATER - water held in saturated material with interstices so small that it is permanently attached to the pore walls, or moves so slowly that it is usually not available as a source of water for pumping.

FIXED MOISTURE - moisture held in the soil below the hygroscopic limit.

flood ice - icing.

fossil ice - ground-ice.

FRAZIL ICE - ice formed by freezing of turbulent water. It is a mush of ice spicules and water resembling slush.

FREE WATER - interstitial gravity water which will freeze at normal temperature (0° C.). According to Bouyoucos it freezes for the first time at the supercooling of -1.5° C.

FRESH-WATER ICE - ice formed by the freezing of fresh water in lakes, streams, or in ground.

FROST-BELT - a ditch that causes an early and rapid freezing of superficial ground forming an obstruction to percolating shallow ground-water.

FROST-BLISTER (Soil blister, Gravel-mound) - a mound or an upwarp of superficial ground caused chiefly by the hydrostatic pressure of ground-water.

FROST-BOIL - accumulation of excess of water at a place of accelerated spring thawing of ground-ice. It usually weakens the surface and may break through causing a quagmire.

FROST-DAM - artificially induced freezing of ground to intercept subsurface seepages which cause icings; equivalent of frost-belt.

FROST-HEAVING - an upward force usually manifested by a more or less marked upwarp due to the swelling of frozen ground.

FROST-MOUND (Ground-ice-mound, ice-mound, earth-mound, gravel-mound (in part), pals, pingo (in part), peat-mound, suffosion conves or suffosion complex, or suffosion knob, hydrolaccolith (in part) - a seasonal upwarp of land surface caused by the combined action of (1) expansion due to the freezing of water, (2) hydrostatic pressure of ground-water, and (3) force of crystallization of ice.

FROST TABLE - a more or less irregular surface that represents the penetration of spring and summer thawing of the seasonal frozen ground (active layer). (Not to be confused with permafrost table.)

FROZEN GROUND - ground that has a temperature 0° C. or lower and generally contains a variable amount of water in the form of ice.

frozen zone - permafrost.

gefronis - permafrost.

GLACIER - A body of ice descending along a mountain valley, commonly commencing as a congealed (recrystallized) mass of snow (firn).

glacier - this term unfortunately is widely used in Alaska to denote ground-ice or sheets of surface ice formed by successive freezing of ground or river seepages which in this report are designated as icings.

GLACIER ICE (Ice of glacial origin) - may be used for ice found under old moraines or outwash deposits.

gravel-mound (Frost blister) - a low mound of earth or sand and gravel formed by hydrostatic pressure and occurring in areas of frozen ground.

GRAVITY WATER (Vadose water) - water in excess of pellicular water and which can, therefore, be drawn away by the force of gravity.

GROUND-ICE (Subsoil-ice, underground-ice, fossil-ice, subterranean-ice, stone-ice, bodeneis, ureis, jordbundsis. Term "glacier" used by miners in Alaska) - bodies of more or less clear ice in frozen ground. Excludes ice of glacial origin.

GROUND-ICE WEDGE - ice wedge.

HYDROLACCOLITH - usually a large frost-mound or an upwarp of ground produced by the freezing of water into a large lenticular body of ice -- in a general way resembling a laccolith. (Pingo).

hydraulics - that part of hydrodynamics, which treats of fluids in motion.

hydrodynamic - pertaining to fluids in motion.

hydrostatic - pertaining to fluids at rest.

HYGROSCOPIC MOISTURE - the thin film of water on the surface of the ground particles which is not capable of movement through gravitational or capillary forces.

ice-field - icing.

ice-heap - icing-mound.

ice-hillock - icing-mound.

ice-mound - frost-mound.

ICE-PIPE - ice wedge of cylindrical shape.

ICE-WEDGE - a narrow crack or fissure of the ground filled with ice which may extend below the permafrost table.

ICING - a mass of surface ice formed during the winter by successive freezing of sheets of water that may seep from the ground, from a river, or from a spring. When the ice is thick and localized it is called icing-mound, and when it survives the summer it is called "taryn".

ICING-MOUND - a localized icing of substantial thickness but of more or less limited areal extent. May also form entirely or in part by the upwarp of a lyer of ice (as in a river) by the hydrostatic pressure of water.

INTRAPERMAFROST WATER - ground-water in unfrozen layers, lenses, or veins within the permafrost.

ISLAND OF TALIK - unfrozen ground beneath the seasonally frozen ground (active layer) surrounded on the sides by the permafrost and extending vertically to the bottom of the permafrost.

ISOPIESTIC LINE - a countour of the pressure surface of an aquifer.

JUVENILE WATER - water which is derived from the interior of the earth and which has not previously existed as atmospheric or surface water.

KARST - uneven topography with short ravines, sink-holes, and caverns, which are produced in a limestone terrain by the solvent action of water.

lamellar ever frozen ground - layered permafrost.

LAYERED PERMAFROST (Layered permanently frozen ground) - ground consisting of permanently frozen layers alternating with unfrozen layers or taliks.

LAYERED PERMANENTLY FROZEN GROUND - layered permafrost.

LEVEL OF ZERO AMPLITUDE - abbreviation of "level of zero annual amplitude".

LEVEL OF ZERO ANNUAL AMPLITUDE - the level to which seasonal change of temperature extends into permafrost. Below this level the temperature gradient of permafrost is more or less stable the year around.

METEORIC WATER - water derived from the atmosphere.

MUCK - mixture of decayed vegetable matter and silt-like material forming the surface layer of the ground in areas of permafrost. Locally, in river valleys, muck may be as much as 100 feet thick.

MUD-POLYGON - polygonal soil, fissure polygon, mud-flat polygon.

NEVE - snow ice.

niggerhead tundra - local name for hummocky tundra in northern Alaska.

OPEN SYSTEM - a condition of freezing of ground when additional supply of water is available either through free percolation or through capillary movement.

PALS - Finnish term for frost-mound or peat-mound. (German plural "Palsen"; Swedish plural "Palsar".)

PASSIVE METHOD (of construction) - method of construction in which the regime of the frozen ground at and near the structure is not disturbed or altered.

PASSIVE PERMAFROST (Passive permanently frozen ground) - permafrost that was formed during earlier colder climates; once destroyed does not appear again.

PEAT-MOUND - frost-mound.

PELLICULAR WATER - water adhering as films to the rock surfaces or to the surfaces of grains that compose the rock. Pellicular water is stored water above the capillary fringe.

PERCOLATION - a type of laminar flow of water in interconnected openings of saturated granular material under hydraulic gradient.

PERELETOK - a frozen layer at the base of active layer which remains unthawed for one or two summers (Russian term meaning "survives over the summer"). Pereletok may easily be mistaken for permafrost.

perennially frozen ground - permafrost.

PERMAFROST (Permanently frozen ground) - a thickness of soil or other surficial deposit or even of bedrock, at a variable depth beneath the surface of the earth in which a temperature below freezing has existed continuously for a long time (from two to tens of thousands of years).

PERMAFROST TABLE (Permanently frozen ground table) - a more or less irregular surface which represents the upper limit of permafrost.

PERMANENT TALIK - a layer of unfrozen ground between the active layer (seasonal frozen ground) and the permafrost (permanently frozen ground) or within the permafrost whose unfrozen state is of many years' duration.

PERMANENTLY FROZEN GROUND - permafrost.

PERMEABILITY - the capacity of water-bearing material to transmit water, (measured by the quantity of water passing through a unit cross section in a unit time under 100 per cent hydraulic gradient).

PERMEABILITY COEFFICIENT - as defined by Meinzer, the rate of flow in gallons a day through a square foot of the cross section of material, under 100 per cent hydraulic gradient, at a temperature of 60° F. In field terms it is expressed as the number of gallons of water per day at 60° F. that is conducted laterally through each mile of the water-bearing bed under investigation (measured at right angles to the direction of percolation) for each foot of thickness of the bed and for each foot per mile of hydraulic gradient.

perpetually frozen soil, subsoil, or ground - permafrost.

PINGO - Eskimo name for "conical hill". Has been used in the past as a local name for a frost-mound. It is suggested that the name pingo should be restricted to frost-mounds that are of longer than seasonal duration and that are, as a rule, of relatively large dimensions.

pingorssarajuk - Eskimo name for a frost-mound "the one that is growing"

pluvoon - (slud, paste). Russian term.

POLYGONAL MARKINGS (Stone-polygons, soil-polygons, mud-polygons, mud-flat-polygons, fissure-polygons (primary and secondary), tundra-polygons, drought-polygons, rudemarks ("rutmarken"), Strukturboden, Polygonboden, Zellenboden, Steinringe, Karreboden, Steinnetz, Spaltennetz, Schuttinseln) - general term for polygonal surface markings of the ground found in the areas that are affected by frost action.

POLYGONAL SOIL - polygonal pattern of the surface of the ground produced by a more or less marked segregation of textural constituents of the ground and also indicated by a slight relief.

POLYN'YA (Russian term) - an unfrozen portion or a window in the river ice which remains unfrozen during all or a part of the winter owing to a local inflow of warm water either from a subaqueous spring or from a tributary.

POROSITY - the property of a rock or soil determined by the presence of interstices of any size or shape, and of any manner of interconnection or arrangement of openings. It is expressed as percentage of total volume occupied by interstices.

PRESSURE SURFACE - the surface to which confined water will rise in non-pumping wells that pierce a common aquifer whose water levels are not affected by a pumping well. (It is a graphic representation of the pressure exerted by confined water on the conduit walls).

PRESSURE-SURFACE MAP - a map showing the contours (isopiestic lines) of the pressure surface of a confined-water system.

PSEUDOISLAND OF TALIK - unfrozen ground beneath the seasonally frozen ground (active layer) surrounded and underlain by continuous permafrost.

RESIDUAL SWELLING - the difference between the original pre-freezing level of the ground and the level reached by the settling after the ground is completely thawed.

SALT WATER ICE - ice formed by the freezing of salt water.

SATIN ICE - acicular ice.

SATURATION, WATER OF - the total water that can be absorbed by water-bearing materials without dilation of the sediments.

SEASONALLY FROZEN GROUND - ground frozen by low seasonal temperatures and remaining frozen only through the winter.

SEEPAGE - the percolation of water through the surface of the earth or through the walls of large openings in it, such as caves or artificial excavations. May be influent seepage (seepage into the ground) and effluent seepage (seepage out of the ground).

SLUD - (Provincial English word for soft, wet mud or mire) ground that behaves as a more or less viscous fluid. It may occur as a superficial deposit or as a layer or lens beneath the surface and may at times be under a considerable hydrostatic pressure. It is "Solifluctional ground" but it is not restricted to the surficial or soil material and its movement is not limited to gravitational flow.

SOIL - the layer or mantle of mixed mineral and organic material penetrated by roots. It includes the surface soil (horizon A), the subsoil (horizon B), and the substratum (horizon C) which is the basal horizon and is limited in depth by root penetration. In engineering practice, under the term soil, are included practically every type of surficial earthen material, including artificial fill, soft shales, and partly cemented sandstones.

soil blister - frost blister.

SOLIFLUCTION - a process of subaerial denudation consisting of the slow gravitational flowing of masses of superficial materials saturated with water.

SPORADIC PERMAFROST - permanently frozen ground occurring as scattered islands in the area of dominantly unfrozen ground.

stable frozen ground - permafrost.

STAMP - a device for determining the strength (or ability) of the ground to support or to withstand a load.

steineis - German for ground-ice.

stone ice - ground-ice.

STONE POLYGON - polygonal areas of fine-texture ground delimited by borders of large stones.

SUBPERMAFROST WATER (subwater) - ground-water in the unfrozen ground beneath the permafrost.

subsoil ice - ground-ice.

subterranean ice - ground-ice.

subwater - subpermafrost water.

suffosion complex - frost-mound.

suffosion convex - frost-mound.

suffosion knobs - frost-mounds.

superwater (suprapermafrost water) - water in the ground above the permafrost.

SUPRAPERMAFROST LAYER - thickness of ground above permafrost consisting of active layer, talik and also the pereletok, wherever present.

SUPRAPERMAFROST WATER - ground-water above the impervious permafrost table.

suprazone - thickness of ground above permafrost consisting of active layer, talik and also the pereletok, wherever present.

SURFICIAL SWELLING - swelling of ground, usually of small magnitude (5 to 10 cm.), caused by the freezing of meteoric waters which penetrate to a small depth below the surface.

SWELLING OF GROUND - increase in volume of surficial deposits due to frost action.

symbiotic method (of construction) - passive method of construction.

taele (Tjäle) - Swedish term for frozen ground.

TALIK - a Russian term for a layer of unfrozen ground between the seasonal frozen ground (active layer) and the permafrost. Also applies to an unfrozen layer within the permafrost as well as to the unfrozen ground beneath the permafrost.

TANGENTIAL ADFREEZING STRENGTH - resistance to the force that is required to shear off an object which is frozen to the ground and to overcome the friction along the plane of contact between the ground and the object.

TARYN - a Siberian term for icings or "ice-fields" which do not melt (thaw) completely during the summer.

TEMPORARY TALIK - a layer of unfrozen ground between the active layer (seasonally frozen ground) and permafrost, whose unfrozen state is due to an occasional warm winter or unusually early snowfall. It usually disappears with the return of the normal winter regime.

THERMOKARST - karst-like topographic features produced by the melting of ground-ice and the subsequent settling or caving of ground.

torfhuegel - peat mound.

TRANSITORY FROZEN GROUND - ground frozen by a sudden drop of temperature and remaining frozen but a short time, usually a matter of hours or days.

UNDERFLOW - movement of ground-water in an underflow conduit.

UNDERFLOW CONDUIT - permeable deposit that underlies a surface streamway and contains ground-water that percolates generally downstream.

underground-ice - ground-ice.

underwater ice - bottom-ice.

ureis - German for ground-ice.

VADOSE WATER - gravity-water.

VOLUME WEIGHT (or VOLUMETRIC WEIGHT) - the ratio of the weight of a unit volume of dry ground to that of an equal volume of water under standard conditions.

WATER-TABLE - in pervious granular material the water-table is the upper surface of the body of free water which completely fills all openings in material sufficiently pervious to permit percolation. In fractured impervious rocks and in solution openings it is the surface at the contact between the water body in the openings and the overlying ground air.

ZERO CURTAIN - a layer of ground between active layer and permafrost where zero temperature (0° C.) lasts a considerable period of time (as long as 115 days a year) during the freezing and thawing of overlying ground.

BIBLIOGRAPHY

BARANOV, I.Ia, Mapping of areas of permafrost: Tr. Kom. po Izuch. Vech. Merzloty, Acad. Sci. USSR., vol. 6, pp. 107-125, 1938. (In Russian with summary in English.)

-- Observations of water freezing: Tr. Kom. po Izuch. Vech. Merzloty, Acad. Sci. USSR., vol. 6, pp. 167-171, 1938. (In Russian with summary in English.)

BARNES, H.T., Ice Engineering, 364 pp., Renouf Publ. Co., Montreal, 1928.

BESKOW, Gunnar, Erdflissen und Strukturboden der Hochgebirge im Lichte der Frosthebung: Geol. För. Förhandl., vol. 52, pp. 622-638, 1930.

-- Tjälbildningen och tjällyftningen: Sveriges Geol. Undersök., ser. C, no. 375, 242 pp., Stockholm, 1935. (In Swedish with summary in English.)

BILIBIN, Iu. A., Active and passive permafrost: Izvest. Russ. Geog. O-va, vol. 69, no. 3, pp. 409-411, 1937. (In Russian.)

BOUYOUCOS, G.J., Classification and measurement of different forms of water in the soil by means of the dilatometer method: Teach. Bull., Mich. Agr. College, no. 36, p. 48, 1917.

-- Movement of soil moisture from small to the large capillaries of the soil upon freezing: Jour. of Agr. Res., 24, no. 5, pp. 427-431, 1923.

BYKOV, N.I., and KAPTEREV, P.N., Vechnaia merzlota i stroitel'stvo na nei /Permafrost and construction on it./, 372 pp., Moscow, 1940.

CRESSY, G.B., Frozen Ground in Siberia: Jour. of Geol., vol. 47, pp. 472-488, 1939.

DATSKY, N.G., Swelling of ground along the railroads under permafrost conditions: Tr. Kom. po Izuch. Vech. Merzloty, Acad. Sci. USSR., vol. 4, pp. 171-187, 1935. (In Russian.)

EAGER, W.L., and PRYOR, W.T., Ice Formation on the Alaska Highway: Public Roads, vol. 24, No. 3, pp. 55-74; 82, Jan.-Feb.-March, 1945.

GERASIMOV, I.P., and MARKOV, K.K., The glacial period in the territory of USSR.: Tr. Inst. Geog., Acad. Sci. USSR., fasc. 33, 462 pp., 1939. (In Russian with summary in English.)

GLAZOV, N.V., Method of studying the degradation of permafrost: Tr. Kom. po Izuch. Vech. Merzloty, Acad. Sci. USSR., vol. 6, pp. 155-161, 1938. (In Russian with summary in English.)

GRIGOR'EV, A.A., Permafrost and Glaciation: Materialy Kom. po Iz. Estestv. Proizvod. Sil USSR., Acad. Sci. USSR., No. 80, Sbornik "Vechnaia Merzlota," pp. 43-104, 1930. (In Russian.)

HAWKES, L., Frost action in superficial deposits, Iceland: Geol. Mag., vol. 61, pp. 509-513, 1924.

HÖGBOM, Bertil, Ueber die geologische Bedeutung des Frostes: Bull. Geol. Inst., Univ. Upsala, vol. 12, pp. 257-389, 1914.

KALITIN, N.N., The role of actinometry in the solution of permafrost problems: Materialy Kom. po Iz. Estestv. Proizvod. Sil USSR., Acad. Sci. USSR., No. 80, Sbornik "Vechnaia Merzlota," pp. 157-176, 1930. (In Russian.)

KOLOSKOV, P.I., On the question of heat melioration in the areas of permafrost and deep winter freezing of soil: Materialy Kom. po Iz. Estestv. Proizvod. Sil USSR., Acad. Sci. USSR., No. 80, Sbornik "Vechnaia Merzlota," pp. 201-231, 1930. (In Russian.)

KORIDALIN, E.A,, Possibility of application of seismic methods to the study of permafrosts: Tr. Kom. po Iz. Vech. Merzloty, Acad. Sci. USSR., vol. 3, pp. 13-19, 1934. (In Russian.)

KUSHEV, S.L., Morphology and genesis of hilly swamps and their geographical distribution: Tr. Kom. po Iz. Vech. Merzloty, Acad. Sci. USSR., vol. 8, pp. 119-161, 1939. (In Russian with summary in English.)

LEFFINGWELL, E. de K., The Canning River Region, Northern Alaska: U.S. Geol. Survey, Prof. Paper 109, 245 pp., 33 figs., 35 pl., maps, 1919.

LUKASHEV, K.I., Mound formation as a manifestation of the tension in permafrost: Ezhegod. Leningrad Univ., no. 10, ser. Geol., Pochvoved. i Geog., vyp. 3, pp. 147-158, 1936. (In Russian.)

-- Granulometric composition of ground in the permafrost area: Ezhegod. Leningrad Univ., no. 16, ser. Geol., Pochvoved., i Geog., vyp. 4, pp. 170-184, 1937. (In Russian.)

-- The region of permafrost. Leningrad Univ., 187 pp., 1938. (In Russian.)

LUKASHEV, K.I. and PANOV, D.G., Organization of permafrost-geological investigations for railroad and building purposes: Tr. Geog.-Ekon. Nauch.-Issled. Inst., Leningrad Univ., vol. 3, pp. 1-34, 1934. (In Russian.)

L'VOV, A.V., Poiski i ispytaniia vodoistochnikov vodosnabzheniia na zapadnoi chasti Amurskoi Zhel. Dorogi/Prospecting for and testing of sources of water supply along the western part of the Amur Railroad/, 881 pp., Irkutsk, 1916. (In Russian.)

MAL'CHENKO, E.V., Permafrost in Eastern Siberia and Yakutiia: Geofiz. Probl. Yakutii, Tr. Kom. po Iz. Yakut. ASSR., Acad. Sci. USSR., vol. 11, pp. 150-176; 223-235, 1928. (In Russian with summary in English.)

-- Climatic conditions in the permafrost region: Materialy Kom. po Iz. Estestv. Proizvod. Sil USSR., Acad. Sci. USSR., no. 80, Sbornik "Vechnaia Merzlota," pp. 105-134, 1930. (In Russian.)

MAKSIMOV, V.M. and TOLSTIKHIN, N.I., On hydrological conditions in the vicinity of the town of Yakutsk: C.R., Acad. Sci. USSR., no. 1, vol. 28, pp. 93-96, 1940. (In English.)

MEL'NIKOV, P.I., Conference on the construction of transport under the permafrost conditions: Vestnik Acad. Sci. USSR., no. 1-2, pp. 103-107, 1940. (In Russian.)

MIDDENDORFF, A. von, Sibirische Reise. Bd. 4, Th. 1, Uebersicht der Natur. Nord. und Ost. Sibiriens, Lief. 3, Klima, pp. 335-523, 1861.

MOTL, C.L., Curing Minnesota frost boils by drains: Eng. News-Record, pp. 270-272, Feb. 12, 1931.

NIKIFOROFF, C.C., On certain dynamic processes in the soils of the permafrost region: Pochvovedenie, no. 2, 1912.

-- The perpetually frozen subsoil of Siberia: Soil Science, vol. 26, no. 1, pp. 61-79, July 1928.

OBRUCHEV, S.V., Solifluctional terraces and their origin, based on studies in the Chukotsk region: Problemy Arktiki, no. 3, pp. 27-48; no. 4, p. 57-83, 1937. (In Russian with summary in English.)

OBRUCHEV, V.A., Engineering work under permafrost conditions: Polevaia Geologiia, 4th ed., vol. 2, pp. 242-250, 1932. (In Russian.)

PETROV, V.G., Measures to protect road construction from destructive effect of icings: Sovetskaia Aziia, no. 3-4, pp. 69-74, 1930. (In Russian.)

-- Naledi /icings/ on the Amur-Yakutsk highway: Acad. Sci. USSR., and Nauch.-Issled. Avto-Dorozh. Inst., Leningrad, 177 pp., album of sketch maps, and 36 pl., 1930. (In Russian.)

-- Experiment to determine the ground water pressure in icings: Tr. Kom po Iz. Vech. Merzloty, Acad. Sci. USSR., vol. 3, pp. 59-72, 1934. (In Russian.)

PETROVSKI, A.A., Electrometric methods to determine the depth of permafrost: Materialy Kom. po Iz. Estestv. Proizvod. Sil USSR., Acad. Sci. USSR., no. 80, Sbornik "Vechnaia Merzlota," pp. 177-184, 1930. (In Russian.)

PODOL'SKI, V., Experience in mine prospecting in the Arctic under the permafrost conditions: Problemy Arktiki, no. 4, pp. 79-82, 1939. (In Russian.)

PONOMAREV, V., Permafrost and mine waters in the Arctic: Sovetskaia Arktika, no. 4, pp. 111-116, 1936. (In Russian.)

PORSILD, A.E., Earth mounds in unglaciated northwestern America: Geog. Review, vol. 28, pp. 46-58, 1938.

POSER, H., Bemerkungen zum Strukturbodenproblem: Centralbl. f. Min., Abt. B., pp. 39-45, 1934.

PRESNIAKOV, E.A., Certain forms of relief due to permafrost: Bull. Russ. Geog. O-va, vol. 69, no. 1, 1937.

ROKHLIN, M., Excavation of prospect pits in the frozen grounds with the use of explosives: Problemy Arktiki, no. 2, pp. 221-223, 1938. (In Russian.)

SHARP, R.P., Ground-ice mounds in tundra: Geog. Review, vol. 32, no. 3, pp. 417-423, 1942.

SHARPE, C.F.S., Landslides and related phenomena; a study of mass-movements of soil and rock. 136 pp., 9 pls., 16 figs. New York, Columbia Univ. Press, 1938.

SHVETSOV, P.F., Permafrost and engineering-geological conditions of the Anadyr' region: Gorno-Geol. Uprav. Glav. Sev. Mor. Put., pp. 1-78, 1938. (In Russian.)

SMITH, P.S., Areal Geology of Alaska: U.S. Geol. Survey, Prof. Paper. 192, pp. 70-71, 1939.

SUSLOV, C.P., Auto-truck road under permafrost conditions in the Enisei forest-tundra: Tr. Kom. po Iz. Vech. Merzloty, Acad. Sci. USSR., vol. 4, pp. 95-145, 1935. (In Russian.)

SUMGIN, M.I., Permafrost in the north of the U.S.S R.: Tr. 1-oi Geol.-Razved. Konf. Glavsevmorputi. Geology and Mineral deposits of the north of U.S.S.R., vol. 3, Vechnaia Merzlota, pp. 7-33, 1936. (In Russian.)

-- Obshchee Merzlotovedenie /General Permafrostology/, Acad. Sci. USSR., 340 pp., 1940. (In Russian.)

SUMGIN, M.I. et al, Instructions for the investigation of permafrost for engineering purposes: Suppl. to Instructions and program directions for the study of frozen and permanently frozen grounds. Acad. Sci. USSR., pp. 253-272, 1938. (In Russian.)

-- Vodosnabzhenie zheleznykh dorog v raionakh vechnoi merzloty /Water supply of railroad in the permafrost areas/. 251 pp. Moscow, Transzheldorizdat., 1939. (In Russian.)

SUMGIN, M.I. and DEMCHINSKI, B., Oblast vechnoi merzloty /The permafrost province/. 238 pp., Moscow, Glavsevmorput' Izd., 1940. (In Russian.)

SVETOZAROV, I., The hydrogeology of permafrost regions, based on investigation in the area of the town Yakutsk: Probl. Sov. Geol., no. 10, pp. 119-132, 1934. (In Russian with summary in English.)

TABER, STEPHEN, Surface heaving caused by segregation of water forming ice crystals: Eng. News-Record, vol. 81, pp. 683-684, 1918.

-- Frost heaving: Jour. of Geol., vol. 37, pp. 428-446, 1929.

-- Freezing and thawing of soils as factor in the destruction of road pavements: Public Roads, no. 6, pp. 113-132, 1930.

-- The mechanics of frost heaving: Jour. of Geol., vol. 38, pp. 303-317, 1930.

TOLSTIKHIN, N.I., Ground water of Trans-Baikalia and their hydrolaccoliths: Tr. Kom. po Iz. Vech. Merzloty, Acad. Sci. USSR., vol. 1, pp. 29-50, 1932. (In Russian.)

-- Underground water in Quaternary deposits in regions of ever-frozen ground: Tr. II Int. Conf. Assoc. Study Quater. Period of Europe, fasc. 2, pp. 56-72, 1933. (In English.)

-- Hydrogeologic conditions of water supply in the areas of frozen zone of the lithosphere (permafrost): Tr. 1-oi Geol-Razv. Konf. Glavsevmorputi. Geology and Mineral Deposits of the North of the USSR., vol. 3, Permafrost, pp. 102-127, 1936. (In Russian.)

-- Mineral water of the frozen zone of the lithosphere: Tr. Kom. po Iz. Vech. Merzloty, Acad. Sci. USSR., vol. 6, pp. 63-78, 1938. (In Russian with summary in English.)

TSYTOVICH, N.A., Principles of construction and design of foundations erected on permafrost: Tr. 1-oi Geol.-Razv. Konf. Glavsevmorputi. Geology and Mineral deposits of the north of the USSR., vol. 3, Permafrost, pp. 78-101, 1936. (In Russian.)

TSYTOVICH, N.A. and SUMGIN, M.I., Principles of mechanics of frozen grounds. Acad. Sci. USSR., 432 pp., 1937. (In Russian; table of contents also in English.)

TYRELL, J.B., Crystophenes or buried sheets of ice in the tundra of North America: Jour. of Geol., vol. 12, pp. 232-236, 1904.

U.S. WAR DEPARTMENT, Aviation Engineers: War Department technical manual, TM 5-255. Washington, D.C., April 15, 1944. 479 pp.

VERCHEBA, A.O., Excavation of test pits in the permafrost: Razvedka Nedr, no. 9/10, pp. 42-43, 1937. (In Russian.)

VOLOGDINA, I.S., The study of the adherence through the congelation forces between the concrete, wood and permafrost: Kom. po Izuch. Vech. Merzl. Acad. Sci. USSR., vyp. 2, pp. 39-83, 1936. (In Russian with summary in English.)

WEINBERG, B.P., et al., Led ("lyod") /Ice/ 524 pp., Moscow, 1940. (In Russian.)

WIMMLER, N.L., Placer-mining methods and costs in Alaska: Bull. Bureau of Mines, no. 259, 236 pp., 1927

WINN, H.F., and RUTLEDGE, P.C., Frost action in highway bases and subgrades Eng. Bull. Purdue Univ., vol. 24, no. 2, pp. 1-100, May, 1940.

YANOVSKI, V.K., On the question of methods of investigation of permafrost for planning of engineering structures: Tr. 1-oi Geol.-Razved. Konf Glavsevmorputi. Geology and Mineral deposits of the north of the USSR., vol. 3, Permafrost, pp. 42-77, 1936. (In Russian.)

ZHUKOV, V.F., Construction of the Yakutsk electric plant foundation on permafrost: Stroitel'naia Promyshlennost', vol. 15, no. 5, pp. 12-1[illegible] 1937. (In Russian.)

F°: 210 200 190 180 170 160 150 140 130 120 110 100 90 80 70 60 50 40 30 20 10 0 10

C°: 100 90 80 70 60 50 40 30 20 10 0 10 20

Conversion Table of Measures

	1	2	3	4	5	6	7	8	9
Cm to in................	0.3937	0.7874	1.1811	1.5748	1.9685	2.3622	2.7559	3.1496	3.5433
Meters (m) to ft..........	3.2808	6.5617	9.8425	13.123	16.404	19.685	22.966	26.247	29.527
Meters to yards..........	1.0936	2.1872	3.2808	4.3744	5.4681	6.5617	7.6553	8.7489	9.8425
Meters to fathoms........	0.5468	1.0936	1.6404	2.1872	2.7340	3.2808	3.8276	4.3744	4.9212
Km to statute miles......	0.6214	1.2427	1.8641	2.4855	3.1069	3.7282	4.3496	4.9710	5.5923
Km to nautical miles.....	0.5396	1.0791	1.6187	2.1583	2.6979	3.2374	3.7770	4.3166	4.8561
Kg to lb [avoir].........	2.2046	4.4092	6.6139	8.8185	11.023	13.228	15.432	17.637	19.842
Kg per m to lb per lin ft..	0.6720	1.3439	2.0159	2.6879	3.3599	4.0318	4.7038	5.3758	6.0477
Calories to B t u.........	3.9683	7.9366	11.905	15.873	19.842	23.810	27.778	31.746	35.715